D0403549

DATE DUE

NO 11 98			

DEMCO 38-296

THE COMMUNICATIONS MIRACLE

The Telecommunication
Pioneers from Morse to the
Information Superhighway

THE COMMUNICATIONS MIRACLE

The Telecommunication
Pioneers from Morse to the
Information Superhighway

JOHN BRAY

PLENUM PRESS • NEW YORK AND LONDON

Riverside Community College
Library
FEB '97 4800 Magnolia Avenue
Riverside, California 92506

TK 139 .B73 1995

Bray, John, 1911-

The communications miracle

Library of Congress Cataloging-in-Publication Data

On file

ISBN 0-306-45042-9

© 1995 Plenum Press, New York
A Division of Plenum Publishing Corporation
233 Spring Street, New York, N.Y. 10013-1578

10 9 8 7 6 5 4 3 2 1

All rights reserved

No part of this book may be reproduced, stored in a retrieval system, or transmitted
in any form or by any means, electronic, mechanical, photocopying, microfilming,
recording, or otherwise, without written permission from the Publisher

Printed in the United States of America

The good effects wrought by founders of cities, law givers, fathers of the peoples, extirpators of tyrants, and heroes of that class, extend but for short times; whereas the work of the Inventor, though a thing of less pomp and show, is felt everywhere and lasts forever.

—Francis Bacon, sixteenth century

Foreword

It has been said that the "electronic revolution" of the past half century constitutes the most important event in human history since the birth of Jesus Christ. Certainly its impact on our daily lives and, indeed, its predication of our future, are so total as to be unquantifiable.

But the subject of this book predates, encompasses, and looks beyond the discovery of the semiconductor and all of its allied technological achievements. In addition, the author has chosen to present his account of the "dramatic advances in telecommunication and broadcasting that have occured in the last century and a half" as an essentially human story. He writes not only about the achievements themselves, but about those who themselves made those giant strides reality. It would be hard to imagine any candidate more qualified to undertake a task as monumental in scale and importance.

Here John Bray has a unique advantage. He was himself a principal player in the subject of his narrative, yet such is the modesty and scholarly discipline of the man that, save for a final chapter of personal reminiscences, not a single sentence of the text is written in the first person—either singular or plural. John Bray, already a world figure in telecommunications, proves himself in this book to be a splendid writer.

I am grateful for the chances of my professional life to have been thrown into close contact—albeit usually all too briefly and often under considerable pressure—with some of the great figures of our time in the boundless realms of science and technology. That one of them would invite me to write a foreword to his work is a compliment of which I am deeply appreciative.

John Bray expresses the hope that this work may inspire young people to follow the path he has trodden with such distinction. I cannot wait to present it to my grandchildren.

Raymond Baxter
Broadcaster and Former Presenter
of TV's *Tomorrow's World*

Preface

This book is an account of the dramatic advances in telecommunications and broadcasting that have occurred in the last century and a half, told through the stories of the people who made it happen. It brings together the lives and achievements of the men whose innovative and creative work in science, engineering, and mathematics from the mid-1800s onwards made possible present-day national and international telecommunications services, sound and television broadcasting.

These major developments can, in many cases, be traced back to specific advances in scientific knowledge, novel inventions, or new system concepts created by certain individuals, whose work can now be seen to have provided the essential keys to progress.

The book explores the background and motivation of each of these men—the *Telecommunicators*—and the social and economic environment in which they worked. It describes their feelings about their discoveries and the significance of their contributions to the modern world.

Few of the technological advances described arose from specifically sponsored/market-oriented cost-effectiveness studies. They were often the result of original creative thought, and sometimes the result of intuition. These were individuals who foresaw a communication need and the scientific and engineering possibilities for meeting it.

The book is a history of technological innovators and innovations in telecommunications and broadcasting that have made possible information technology that is taken for granted in the modern world. It foresees many more changes that will have a profound effect on the way people live and work in the future.

OUT OF THE BLUE?

Two books that were part of the author's reading in "light current" electrical engineering at Imperial College, London, in the 1930s were Rollo Appleyard's *Pioneers of Electrical Communication*, published in 1930, and E. T. Bell's *Men of Mathematics*, published in 1937. By their approach these books made the reader aware of the reality of the innovative process, and made it clear that radically new ideas and technological advances do not appear "out of the blue."

More often than not, these ideas are the product of concentrated and continuing thought in a chosen field, allied to intuition, and leading to a quantum jump forward in understanding. They led at least one reader to pursue a professional career as near as possible to the "sharp cutting edge" of technological advance in telecommunications.*

Since the 1930s the development of telecommunications and broadcasting has continued at an accelerating pace, with many innovative and clearly defined steps along the way. In many cases the pioneers responsible can be readily identified—generally by consensus judgment, for example by scientific and engineering institutions. Other cases may be less obvious and, in a commercially competitive world, may need to be selected for the purposes of this book on a more personal and subjective basis, for which the writer makes apology in advance where less than justice may have been done to other claimants for "pioneer" status.

Clearly it is not possible in a single book covering such a broad field to follow in detail the ongoing development beyond the first highly innovative step, but an endeavor has been made in each case to indicate

---

*In the author's case, as Director of Research of the Post Office, now British Telecom, from 1966 to 1975 at the Dollis Hill, London, and Martlesham, Suffolk, laboratories.

in broad terms why the selection has been made and justified in terms of its long-term impact.

Innovation in telecommunication and broadcasting technology is not a national monopoly; the ferment of new ideas exists in many organizations and in many countries, and sometimes in the minds of lone individuals. While in this book some prominence is given to British innovators and inventors, it is clear that many key ideas originated elsewhere, notably in the United States, Germany, France, and Austria, and more recently in Japan. It is hoped that the more important of these have been identified and described.

Engineers, scientists, and mathematicians in the Bell Telephone Laboratories, USA, have made many important contributions to the advance of telecommunications, and it has been the writer's privilege to have been associated with some of these, from the development of transatlantic shortwave radio communication in the 1930s to cooperation in the pioneering TELSTAR communication satellite project of the 1960s. These contacts were enhanced, and given a more personal basis, by a Commonwealth Fund (Harkness Foundation) Fellowship spent at the Bell Laboratories at Murray Hill and Holmdel during 1955–1956.

AIMS OF THE BOOK

It is hoped that this book will provide a human thread through the history of technological advances in telecommunications and broadcasting that will inspire recent graduates and those studying in polytechnics and universities to seek a rewarding career in one of the most important developments of our time, and one in which there is more to be achieved. The book is also aimed at the innovators of tomorrow in the hope that the innovators of yesterday provide inspiration to them to match the communication needs of the future.

No apology is made for concentrating on technological innovation in this book; without the supporting technology none of the far-reaching services now available in telecommunications and broadcasting would have been possible.

Nevertheless it is clear that innovations in design, manufacture, finance, and exploitation also have important roles to play in bringing the results of technological innovation to fruition, and these too create new career possibilities.

Acknowledgments

The writer thanks the authors, editors, and publishers of the many books that have been found useful in preparing the text, and to which reference has been made in the bibliography at the end of each chapter.

Certain books—notably W. J. Baker's *History of the Marconi Company*, *History of Engineering and Science in the Bell System: Transmission Technology (1925–1975)*, edited by Dr. Gene O'Neill, and George Brown's *And Part of Which I Was: Recollections of a Research Engineer*, which gives a colorful insight into the development of color television in the United States—have been particularly useful. Personal records of Bell Laboratories people—such as H. T. Friis' *75 Years in an Exciting World*, J. R. Pierce's *The Beginnings of Satellite Communication*, and R. Kompfner's *The Invention of the Travelling-Wave Tube*, which had an important role in microwave radio-relay and satellite systems—have contributed to the innovation story.

Bell Laboratories Record has provided a very readable account of developments in that organization. In the United Kingdom the *Post Office Electrical Engineers' Journal*, the *Journal of British Telecommunications Engineering*, and the *Proceedings of the Institution of Electrical Engineers, London* have provided valuable documentation.

The International Telecommunication Union publication *From Semaphore to Satellite,* published on its centenary anniversary in 1965, has provided a useful picture of telecommunications development to that date. In the specialized switching field, Robert Chapuis' *100 Years of Telephone Switching: 1878–1978* gives a comprehensive account for that period.

The author thanks former colleagues in the Engineering Department of the British Post Office (now British Telecom), including Charles Hughes, Michael Holmes, Roy Harris, J. Martin, M. B. Williams, and Murray Laver, for their always helpful suggestions and to Dr. J. R. Tillman, former Deputy Director of Research PO/BT, for his valuable comments on the text and additional material.

He is particularly grateful to Dr. Gene O'Neill, formerly of AT&T Bell Laboratories, for encouragement during the writing of the book and constructive reviewing of the text.

Various firms and organizations, including the Marconi Co.; Standard Telephones and Cables Ltd.; Corning Glass, USA; AT&T and Bell Laboratories, USA; The Royal Television Society, UK; the David Sarnoff Research Center Archives, USA; the British Broadcasting Corporation, and the Independent Broadcasting Authority, have provided copies of documents and photographs. The U.S. National Academy of Engineering has supplied information on the Draper Award made to the coinventors of the microchip. The Library of the Institution of Electrical Engineers, UK, has been particularly helpful in providing loan copies of relevant books.

Several organizations have provided photographs and illustrations of historical and technical interest, including:

Hunterian Art Gallery, University of Glasgow (Lord Kelvin)
National Portrait Gallery (Lord Rutherford)
IEE, London (Sir J. J. Thomson)
U.S. Library of Congress (Alexander Graham Bell)
Marconi Co. Archives (Marconi and equipment, Writtle Tr.)
BBC (Capt. P. P. Eckersley and broadcasting pioneers)
BT Archives (PO cable ship, BT Tower, System X equipment)
BT Educational Dept. (telephone pioneers)
Science Museum, London (telegraph cable map, microchip)
Northern Telecom (traveling-wave tube)
Texas Instruments (LS integrated circuit)

Muirhead Systems (facsimile equipment)

AT&T, USA (various illustrations from Bell Laboratories' *Record* and the *History of Engineering and Science in the Bell System*)

Thanks are also due to Mr. Iain Vallance, Chairman of British Telecom Board, and Dr. Alan Rudge, BT Managing Director Development and Procurement, for making available the BT Technical Library, Martlesham, BT Archives London, and also the services of the BT Design Centre, Martlesham in preparing illustrations for publication.

Contents

1

Introduction

TODAY'S WORLD OF TELECOMMUNICATIONS AND BROADCASTING

The present-day world telecommunication network is the most complex, extensive, and costly of mankind's technological creations and, it could well be claimed, the most useful.

Together with sound and television broadcasting, telecommunication provides the nervous system essential for the social, economic, and political development of civilization. It enables any user of the 700 million or more world total of telephones in homes and offices, ships, aircraft, and cars to communicate with any other, regardless of distance.

It provides means for the fast distribution of documents and the "electronic" letter mail; it enables people to communicate by data transmission over the network with remote computers and data banks, and computer with computer.

The long-distance transmission of television signals is now commonplace, even from far-distant planets of the solar system. The use of television for business conferences between offices in distant locations is increasing as more wide-bandwidth channels become available in the world telecommunication network.

And a whole new art of information access, exchange, and processing—information technology (IT)—is developing whereby users in home and office can gain access to virtually unlimited pages of information, visually displayed, from local or remote data banks via the telecommunication network, and interact with the displayed information or send messages on a person-to-person basis.

The future impact of information technology linked to telecommunications could well be immense: (1) by removing the need to travel to communicate, the vast waste of human and material resources needed to provide ever-expanding rail and road facilities for countless millions of commuters every day from homes to city offices could be minimized; and (2) by diverting much office work to villages and small towns, the quality of life could be enhanced and the rural economy sustained.

This vast development in scale and range of telecommunication services and facilities has become possible mainly by a remarkable and continuous evolution in telecommunication system concepts and the supporting technology, much of which has taken place in the last two or three decades, but of which the mathematical, scientific, and conceptual origins can be traced back a century or more.

As to the advances telecommunications technology has made, one has only to compare the primitive slow-speed telegraphs, the pole-mounted copper wires, and the manual telephone exchanges of the early 1900s—then the only means of intercity communication—with the advanced coaxial cable, microwave radio-relay, optical fiber, and Earth-orbiting-satellite transmission systems, and the fast computer-controlled electronic exchange switching systems handling voice, data, and vision, in use today.

Device technology has made massive leaps forward, from the thermionic valves of the early 1900s to the revolutionary invention of the solid-state transistor and microtechnology that puts thousands, even millions, of transistors and related devices on a single match-head-size microchip.

These advances in technology have not only made possible a vast growth in the scale and variety of telecommunication services, they have enabled the cost to the users to be progressively reduced in real terms.

And as the distances over which signals need to be transmitted have increased, so new techniques such as pulse-code modulation and digital transmission have been evolved which greatly enhanced quality, flexibility, and reliability.

The technology of broadcasting has also evolved from the monaural amplitude-modulation long- and medium-wave sound broadcasting services introduced in the early 1900s to the stereophonic high-quality frequency-modulation VHF sound broadcasting and high-definition color television UHF services, with teletext (visually displayed alphanumeric information) capability, in the later 1900s. And television broadcasting direct from satellites to the home is now here. These developments have not come about fortuitously; rather, they have been initiated by the highly creative and original thinking of a number of individuals over a century or more.

It is the purpose of this book to seek to identify at least some of the scientists, engineers, and mathematicians whose contributions can each be seen, in retrospect, to have provided an essential key or stimulus to the major developments that followed in later years, and to assess the significance of their individual contributions. However, while some of these innovators may have been inspired by visions of the future, it is certain that very few if any could have foreseen the immense impact that their work has had, and will continue to have, on telecommunications in the 20th century and beyond.

In selecting those on whose work attention has been focused, it has to be recognized that one may be doing less than justice to others who were providing the background from which those creative ideas grew, and those who pursued those ideas to practicality and commercial success. And of course, judgment as to what is a major and specially important contribution may itself be highly subjective.

Nevertheless, looking back over the long history of telecommunications development, it would seem that there are real peaks of creative achievement as clearly recognizable as Darwin's in the theory of evolution and Newton's in mathematical physics and astronomy.

It is as a tribute to these men, and in the hope that it may, in some degree, inspire a new generation of "telecommunicators" to further achievement, that this book has been written.

INNOVATION, INVENTION, AND COMMUNICATION

As Edward de Bono has observed, "There is nothing more important than an idea in the mind of man—Man's achievements are based on man's ideas."[1]

The history of technology has demonstrated repeatedly that it is the innovative and creative idea in the mind of an individual—sometimes stimulated by others working in the same field but often working alone—that has provided the start of developments that can now be seen to have been of major importance in the onward march of civilization.

Most innovations and inventions in telecommunications and broadcasting have originated from a combination of need, e.g., to communicate person to person at a distance or to a mass audience, of recognition of the relevant scientific or engineering principles and possibilities, and at times chance observation of phenomena.

Inventions, and inventors, have not always been welcomed or understood by those in authority. For example, the Italian government showed little interest in Marconi's work on wireless, the British Post Office initially failed to recognize the value of Strowger's invention of the automatic telephone exchange, and even the transistor was regarded by one leading scientist as "unlikely to replace the thermionic valve."

It may be that, as the scale of telecommunications and broadcasting has expanded to cover the globe, the scope for individual contributions to technological development has been overlaid by teamwork and the massive power of the large corporations; nevertheless the role played by individuals remains crucial and deserves to be recognized.

One message from this survey of innovation and invention stands out with clarity: no worthwhile invention was ever created because a team of cost-study experts and accountants said "you must invent this or that!"

No apology is made for including broadcasting as well as telecommunications in this survey; many of the scientific discoveries, mathematical studies, and technological developments that made advances in telecommunications possible have also contributed to the evolution of broadcasting—from the discovery of radio waves to the invention of the transistor and the microchip.

REFERENCE

1. *Eureka: An Illustrated History of Invention,* edited by Edward de Bono (Thames and Hudson Ltd., Cambridge, 1974).

2

Creators of the Mathematical and Scientific Foundations

Telecommunications is an exact science that depends heavily on scientific discoveries and advances in mathematics made by a small number of scientists and mathematicians in Europe during the mid-19th century, among whose names those of Volta, Ampère and Ohm, Faraday and Oersted, Maxwell, Hertz, and Heaviside are outstanding.

In much of their work, these men were exploring hitherto uncharted areas of knowledge and grappling with difficult concepts as to the nature of electricity and magnetism. They sought and found means for defining, and expressing in quantitative terms, the various electrical and magnetic phenomena they encountered and evolved new terms such as voltage, current, impedance, lines of force, field strength, and electromagnetic waves. By so doing the prediction of performance of electromagnetic systems became possible and the bases for electrical design were laid.

The lives and work of these men have been described in detail and in depth in Rollo Appleyard's classic book *Pioneers of Electrical Communication*,[1] published in 1930, but now out of print. Much of the account in this chapter of the early pioneer scientists and mathematicians is based on Appleyard's book. Use has also been made of the summaries of the work of the pioneers prepared by the British Telecom Education Service.[2]

ALESSANDRO VOLTA (1745–1827)

Alessandro Volta has a lasting claim to fame: the electrical unit of potential, the "volt," was named for him.

Volta was born in 1745, the son of a well-to-do merchant of Venice. His education was mainly at the Royal Seminary in Como, where he had leanings toward poetry. However, a strong interest in chemistry and electricity took over and by 1779 he had been elected to the Chair of Physics at Padua University.

He traveled extensively, visiting England, France, and Germany; this was remarkable since the Napoleonic Wars were in progress and Europe was in a turmoil. Many of his contacts in these travels were with philosophers and leading men of science, e.g., Voltaire, Priestley, Lavoisier, and Laplace.

He was also acquainted with Luigi Galvani of Bologna and knew of Galvani's discovery that a frog's muscles twitched when joined by a pair of dissimilar metals. From this chance observation grew Volta's most significant invention, the "voltaic pile"—a series connection of pairs of dissimilar metals separated by membranes moistened with acid—that constituted an electric battery.

This invention was described in a paper "On the Electricity Excited by the Mere Contact of Conducting Substances of Different Kinds," recorded in the *Philosophical Transactions of the Royal Society, London* in 1800.

Although Volta produced a number of other electrical inventions, e.g., the "electrophorus," which generated static electricity by friction on an insulator, and a "condensing electroscope" for detecting small charges of static electricity, there is little doubt that the voltaic pile—the forerunner of the modern multicell battery—was by far the most valuable.

Figure 2.1. Portraits of Volta (top left), Ohm (top right), and Ampère (bottom center).

ANDRÉ-MARIE AMPÈRE (1775–1836)

The name of this distinguished French scientist has been honored for all time by designating the unit of electrical current as the "ampere." He was born in 1775 when France was moving toward the Revolution that began in 1789; his father died on the scaffold and Ampère's early life was much clouded by this traumatic event.

He sought solace and became interested in mathematics and studied at the École Centrale in Bourg, where he set up a laboratory for the study of the physical sciences. In 1804 he was awarded a professorship at the École Polytechnique in Paris. His scientific work covered a broad front, and he made many original contributions in mathematics, mechanics, electricity, magnetism, optics, and chemistry.

His most significant contribution in electricity and magnetism concerned the mechanical forces between current-carrying conductors, and the formulation of the laws governing such forces, described in a presentation to the Académie des Sciences in 1820. In so doing he was building onto Oersted's observations on the deflection of a magnetic needle by a current-carrying conductor.

A useful by-product of Ampère's work in electromagnetism was his "astatic galvanometer," in which the deflecting coils were so arranged as to neutralize the effects of external magnetic fields such as the Earth's, thereby increasing the usable sensitivity.

Ampère's formulation of the laws relating to the forces generated between current-carrying conductors later proved to be of major importance in the design of telegraph apparatus, telephone receivers, and loudspeakers.

GEORG SIMON OHM (1789–1854)

As for Volta and Ampère, Ohm's name has by an international agreement been used to designate an important electrical unit, in his case that of resistance—the "ohm."

According to Ohm's biographer Rollo Appleyard:

> A century ago (early 1830s), the science and practice of electrical measurement and the principles of design for electrical instruments hardly existed. With a few exceptions, ill-defined expressions relating to quantity and inten-

sity, combined with immature ideas of conductivity and derived circuits, retarded the progress of quantitative electrical investigations. Yet, amidst this confusion, a discovery had been made that was destined to convert order out of chaos, to convert electrical measurement into the most precise of all physical operations, and to aid almost every other branch of quantitative research. This discovery resulted from the arduous labours of Georg Simon Ohm.

Ohm, who was to achieve fame as a physicist, was born in 1789 in the university town of Erlangen, in Bavaria; a brother, Martin Ohm, was born in 1792 and became a noted mathematician.

Georg Ohm studied mathematics and physics at Erlangen University; however, for reasons of economy, he also had to teach, which at that time he found irksome.

In the early 1800s, conditions in Bavaria became difficult as the struggle against Napoleon developed. Ohm left his native Bavaria in 1817 to settle in Cologne, where he obtained a Readership in the University, and found both freedom and appreciation.

He had developed a growing interest in studying the conductivity of metals and the behavior of electrical circuits and, to pursue these studies more effectively, he gave up teaching in Cologne and went to live in his brother's home in Berlin. Another objective was to prepare a thesis that could be of help in achieving a University professorship.

After intensive studies he wrote *Die galvanische Kette, mathematisch bearbeitet*, a book that established for the first time the relationship between voltage (potential), current, and resistance in an electrical circuit, immortalized in the simple relationship $I = E/R$.

Of considerable importance was the concept—due to Ohm—that the effective potential in a circuit was the sum of the individual potentials, and the effective resistance was the sum of the resistances of the component parts of the circuit.

There was for a time opposition to Ohm's law by one or two physicists, and by the philosopher Hegel who disputed the validity of Ohm's experimental approach. Eventually Ohm's triumph came in 1841 when the Royal Society of London awarded him the Copley Medal for his work.

Recognition followed in Bavaria and Germany, by numerous professional and academic appointments, including one as adviser to the State on the development of telegraphy.

The applicability of Ohm's law to electrolytes and thermoelectric junctions, as well as to metallic conductors, was in time recognized, and it remains today the most widely used and valued of all of the laws governing the behavior of electrical circuits.

HANS CHRISTIAN OERSTED (1777–1851)

Oersted's major contribution to electrical science was the demonstration—from the simple observation of the deflection of a magnetic compass needle—that a magnetic field existed around a current-carrying conductor. This experimental proof of the close connection between electricity and magnetism stimulated studies in depth by many other scientists, including Faraday, Maxwell, and Heaviside; in recognition of his work the unit of magnetic field strength was named the "oersted."

He was born in 1777 at Rudkjobing, a small town on the Danish island of Langeland, where his father was an apothecary. Educational facilities on the island were limited; Hans and his young brother Anders received their primary education largely through the diligence and encouragement of their parents. Through their hard work the brothers gained places at Copenhagen University where Hans was awarded a Ph.D. in 1797.

His interests were initially in literature and philosophy, and included philosophical studies seeking a unity between natural phenomena such as light, heat, chemical action, electricity, and magnetism.

Volta's invention of the voltaic pile, which provided for the first time a continuous source of electrical current, inspired Oersted to concentrate on physics and in 1806 he was appointed Professor of Physics at the University of Copenhagen. He traveled extensively in Germany and France, discussing experimental philosophy with men of science and forming lifelong friendships. One was with Johann Wilhelm Ritter (1776–1810) who had invented a battery (or secondary cell) which could be recharged repeatedly from a voltaic pile.

The claim that a magnetic compass needle could be deflected when a current-carrying conductor was in its vicinity was first reported in 1820, following an almost chance observation in the course of a classroom lecture. What is important, however, is that Oersted's earlier philosophical studies on the essential unity of natural phenomena such

Figure 2.2. Hans Christian Oersted and his compass needle experiment demonstrating the magnetic field around a current-carrying conductor.

as electricity and magnetism made him realize immediately the profound significance of this seemingly chance observation.

Oersted wrote an account in Latin—then the normal language for communicating scientific discoveries and theories—of his compass needle experiment and his conviction that every voltaic circuit carries with it a magnetic field, and sent copies to Societies and Academies in all of the capitals of Europe. He received many honors in recognition of his work, including—like Ohm—the Copley Medal of the Royal Society in London.

While Oersted's discovery had an early application to the electric telegraph—as for example Wheatstone's needle telegraph—perhaps the most valuable outcome was the stimulus it gave to the mid-19th-century scientists, physicists, and mathematicians who were seeking a unified electromagnetic theory.

MICHAEL FARADAY (1791–1867)

Michael Faraday's discoveries in electromagnetism provided the scientific basis for devices that were to become the essential keys to a wide range of new industries, ranging from induction coils and transformers to motors and dynamos essential for the electrical power generation, distribution, and utilization industries, as well as telecommunications. His name has been immortalized in the electrical unit of capacitance—the "farad."[3]

Michael Faraday was born in London in 1791; his early education was in a common day school, amounting to little more than the elements of reading, writing, and arithmetic. At the age of 13 he became an errand boy to a bookseller, which enabled him to learn bookbinding. But more importantly it gave him an opportunity to read extensively, reading that included the articles on electricity in the *Encyclopaedia Britannica*.

When he was 21 he attended Sir Humphry Davy's lectures on heat, light, and the relationship between electricity and chemical action, at the Royal Institution in London. He became Davy's assistant and learned much from him, including the preparation of his own scientific papers and the art of public lecturing.

Faraday's interest in electromagnetism was stimulated by news of Oersted's compass needle experiment, which revealed that there was a magnetic field around a current-carrying conductor which could deflect a magnetic needle. From this Faraday deduced the principle of, and demonstrated in 1821, a simple electric motor in which a wire carrying a current rotated around a fixed electromagnet.

He was elected a Fellow of the Royal Institution in 1824, but not without some opposition from his mentor Sir Humphry Davy.

By 1826 he again turned his attention to electromagnetism, studying the work of Oersted, Ampère, and Ohm. He reasoned that

> If an electric current in a conductor produces a magnetic condition, why should not a conductor near a magnet show an electric condition?

This line of thought led Faraday in 1831 to a series of experiments that were later to have remarkable results. He gave much thought to what was happening in the field around magnets and current-carrying conductors, visualizing "lines of force" that linked opposite magnetic poles, or electric charges, and which exercised a pull or tension along such lines. This concept later enabled mathematical physicists such as

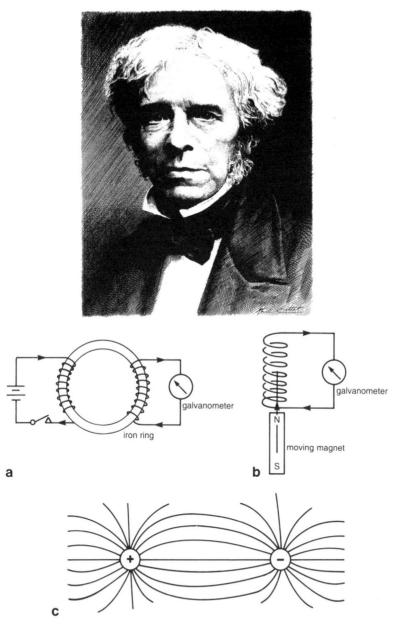

Figure 2.3. Michael Faraday and his electromagnetic experiments demonstrating (a) electric induction, (b) magnetoelectric induction, and (c) lines of force between oppositely charged poles.

Maxwell to give quantitative expression to electromagnetic phenomena.
From his experiments Faraday deduced the fundamental law of
electromagnetism:

> Whenever a closed conductor moves near a magnet in such a manner as to
> cut across the magnetic lines, an electric current flows in the conductor.

His experiments led Faraday to describe two groups of electromagnetic phenomena:

1. Those in which a varying current in one coil produces transient
   currents in a second coil. This he called "electric induction." It
   was the basis of induction coils used later, for example by Edison
   in the telephone, and the alternating current transformers used
   in the telecommunications and electrical power industries.
2. Those involving relative motion between a magnet and a conductor. This he called "magneto-electric induction." It was the
   basis of, for example, the telephone receiver in telecommunications, and the loudspeaker in broadcasting; it gave birth to electric motors and dynamos for the electrical power industry.

Faraday later turned his attention to the decomposition of liquids
by electricity, creating new terms such as *electrolyte* and *electrolysis, anode*
and *cathode*; the latter were subsequently to have wide currency in the
thermionic valves used in telecommunications.

In 1855, when he was 64 years old, Faraday completed his massive
book *Experimental Researches,* embodying the experiments, theories, and
results derived from 24 years of intensive research and study.

In 1867 Faraday died peacefully in his study in a "grace and favour"
house at Hampton Court given him by Queen Victoria in recognition of
his scientific achievements. He was a remarkably modest man, refusing
many honors and financial rewards for his work. Among his many
friends were Wheatstone and Cooke, who later achieved fame for their
contributions to the invention of the electric telegraph.

His happy marriage and profound religious convictions gave him
a serenity of mind that must have helped greatly in his scientific work.
However, he is said not to have thought of himself as a scientist, but
rather a philosopher, i.e., a student of the phenomena of nature.

OLIVER HEAVISIDE (1850–1925)

Oliver Heaviside's contributions to electromagnetic theory are now recognized as major advances in understanding the principles of operation and design of long telegraph and telephone cables, but in their day they were considered as most abstruse and beyond the understanding of ordinary mortals. His work provided, for the first time, an accurate basis for determining the maximum speed of signaling on telegraph cables and the conditions for achieving a "distortionless" cable; his studies of the transmission of telephony on long cables revealed the possibility of improved speech quality by "loading" the cable, i.e., adding inductance at intervals. And his work on electromagnetic wave propagation revealed the possibility of round-the-world transmission by waves guided between the earth or sea and an electrified layer in the upper atmosphere, now known as the "Heaviside layer."

To achieve his objectives, Heaviside had to invent highly unconventional mathematical methods and symbolism, including the use of "operators," much of which he arrived at intuitively and to which

Figure 2.4. Oliver Heaviside (left) and Michael Idvorsky Pupin.

the mathematical purists objected for alleged lack of rigor. Even his contemporaries working on electromagnetism found his work difficult to understand; the editors of the learned society journals to whom he looked to publish his work found him very awkward to deal with because of his unwillingness to adapt his manuscripts to the needs of their readers.

During a half century of intensive theoretical work, Heaviside wrote a great deal. Much of it is published in his *Electrical Papers* (two volumes) and *Electromagnetic Theory* (three volumes); other material has been preserved in the manuscript collection held at the Institution of Electrical Engineers, London, together with letters to his contemporaries, including Maxwell, Faraday, Lord Rayleigh, and Professor G. F. FitzGerald (Dublin). FitzGerald wrote in a review of Heaviside's *Electrical Papers:*

> They teach a sound theory of telegraphs and telephones ... which there is every prospect may lead to vast improvements in telegraphy and may even make it possible to work a telephone across the Atlantic.

Heaviside's theoretical prediction of the value of inductive loading on long cables was further developed and demonstrated practically by Michael Idvorsky Pupin (1858–1933). Pupin was born in the former Yugoslavia in 1858; he arrived in the United States as a penniless immigrant and later graduated from Columbia University. After studying in Germany under Helmholtz and Kirchhoff, Pupin returned to New York and became Professor of Electro-Mechanics at Columbia University.

It is interesting to note that, prior to the outbreak of the Second World War in 1939 and before deep-sea repeaters were practicable, the Bell Telephone Laboratories in the United States had designed an inductively loaded and repeater-less transatlantic cable providing a single telephone circuit. Implementation of the project was postponed by the War, and later replaced by a cable with valve-operated repeater/amplifiers providing 36 telephone circuits.

Heaviside was born in Camden, London, in 1850; details of his early education are lacking. After leaving school he joined the Great Northern Telegraph Company, perhaps through the interest of his uncle Sir Charles Wheatstone, and was also employed by the Anglo-Danish Cable Company. An elder brother, Arthur West Heaviside, was a Divisional Engineer of the British General Post Office.

Oliver and Arthur shared an interest in telegraphy and together carried out a number of experiments. A memorandum of 1873 records how they demonstrated "duplex telegraphy," i.e., simultaneous both-way transmission, over an artificial line with a needle telegraph. Later they demonstrated duplex telegraphy over an actual line between New-castle and Sunderland. Their experiments at this time included work on microphones which perhaps later stimulated Oliver to include tele-phony in his theoretical studies.

In 1882 Heaviside began his famous series of papers in the *Electri-cian* on the theory of propagation of electromagnetic waves on conduc-tors and in space. This work is notable, inter alia, for its establishment of the principle of duality between electricity and magnetism.

Hertz's discovery in 1886 of the wave character of electromagnetic radiation stimulated Heaviside to further theoretical studies and by 1888 he was contributing to the *Philosophical Magazine* on this subject.

In 1889 Oliver and his parents moved to Paignton in Devonshire, a move that perhaps isolated him from experimental work. This, his parents' death, and his own increasing deafness made him more and more a recluse, concentrating increasingly on his theoretical studies. His working habits became more strange, involving long hours, often into the early morning, spent in a closed smoke-filled room.

He was autocratic by nature, highly intolerant of poseurs and those who failed to understand his work. But he had superb powers of concen-tration, a remarkable intuitive perception of the nature of electromagne-tism, and an ability to translate this into theoretical statements of im-mense value for quantitative design.

He died in relative poverty in Torquay, at the age of 75. He was a member of the Society of Telegraph Engineers, a forerunner of the present Institution of Electrical Engineers, London, of which he was made an honorary member. Fittingly, he was awarded the IEE's first Faraday Medal. To him the science of telecommunication owes a massive debt, unrewarded in his lifetime.

HEINRICH RUDOLF HERTZ (1857–1894)

It can be said that the work of the earliest pioneers of electromag-netic studies culminated in 1886 in Karlsruhe when Heinrich Hertz

Figure 2.5. Heinrich Rudolf Hertz (left) and James Clerk Maxwell.

demonstrated experimentally the wave character of electrical transmission in space, as well as along wires. However, Hertz himself was the first to acknowledge that his work was significant because it verified the theoretical predictions made by others, including Maxwell, Faraday, and Helmholtz.

In a sense it is fortunate as well as fortuitous that Hertz's experimental work, because of the convenient dimensions of his laboratory apparatus, involved centimetric-wavelength radio waves. These could be readily reflected by sheets of metal, focused by parabolic mirrors, and deflected by dielectric prisms. His work thus had application not only to the propagation of long- and medium-wavelength radio waves first exploited by Marconi, it later prepared the way for microwave radio-relay systems, radar, and satellite communications. In honor of Hertz's work the unit of frequency—the cycle per second—has been internationally named the "hertz."

Hertz was born in Hamburg in 1857, the eldest son of a Jewish advocate and a member of a prosperous family of middle-class merchants. His grandfather studied natural science as a hobby and left his grandson a small laboratory where the young Hertz carried out simple experiments in physics and chemistry. He early demonstrated an ability in languages, and became proficient in English, French, and Italian—a skill that must have helped him to understand the work of his contemporaries in other countries.

His most formative studies were at the University of Berlin under Helmholtz and Kirchhoff, where he carried out individual research on a largely philosophical question "concerning the extra current manifested when an electric current starts or stops"—a problem also studied by Maxwell. The University awarded him a doctorate for his thesis, with the added and unusual distinction of "magna cum laude." In 1880 he became a demonstrator in physics at the University of Berlin under the direction of Helmholtz, leaving in 1883 to take up a professorial post at the University of Kiel, where he studied intensively electromagnetism, with particular attention to Maxwell's work in this area.

In 1885 he became Professor of Experimental Physics at the Technische Hochscule in Karlsruhe. It was here that, in 1886, he noted that the discharge of a Leyden jar (a charged capacitor) through one of a pair of spiral coils caused a spark to pass across a short air gap between the ends of the other coil.

It had been pointed out by Oliver Heaviside, in the *Philosophical Magazine,* 1877, that the oscillatory nature of the discharge of a charged capacitor in association with a self-inductance was first disclosed by Joseph Henry in 1842. Sir William Thomson (later Lord Kelvin) had also studied the theory of this phenomenon, reported in the *Philosophical Magazine,* 1853, and experimental proof had been shown by Bezold in 1870. Although Hertz was unaware of Bezold's work at the time of his first experiments, he was the first to acknowledge this when reporting his own discoveries.

An important step forward was made when Hertz demonstrated the principle of resonance between the transmitting and receiving circuits—initially by dispensing with the Leyden jar in his spiral coil experiments—which increased the distance over which signals could be detected by making these circuits identical.

A further step was to realize that the detection distance was more favorable than the Newtonian law of inverse squares, in fact more nearly an inverse distance relationship.

His invention of the "linear" oscillator—two straight metal rods terminated by metal spheres providing capacitance—was the forerunner of dipole aerials used in present-day radio, radar, and television systems. With it he was able to demonstrate that the waves produced were

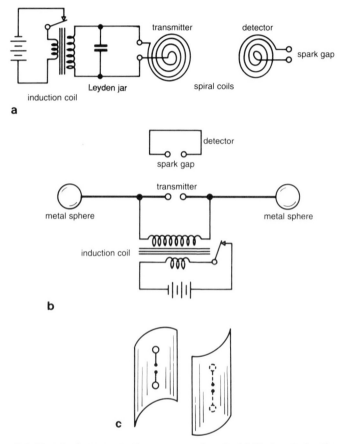

Figure 2.6. Hertz's electromagnetic wave experiments: (a) Hertz spiral coil experiment; (b) Hertz linear oscillator; (c) parabolic reflectors.

"linearly polarized," i.e., the oscillations of the electric component of the radiated field were parallel to the rods of the oscillator. A receiver with rods at right angles to the transmitting rods would pick up virtually no signal. This principle is valuable for the avoidance of interference between radio systems sharing the same frequency, thereby aiding economy in spectrum utilization.

   The demonstration that short-wavelength radio waves could be efficiently concentrated into beams by parabolic reflectors of dimensions comparable with, or greater than, the wavelength later led to directional aerials providing power gain, i.e., more effective transmission and longer range than is possible between dipole aerials by themselves. The radio frequencies used in Hertz's experiments were on the order of 100 MHz, the wavelength being about 30 cm and the parabolic reflectors about 100 cm across.

Hertz's discoveries and his studies of their relationship to Maxwell's theories were published in 1891 in a treatise written in German and dedicated to Helmholtz; this was translated into English by Professor D. E. Jones, with a preface by Lord Kelvin, under the title of *Electric Waves, Being Researches on the Propagation of Waves of Electric Action with Finite Velocity through Space.*

The reference to "finite velocity" in this title is interesting since it is in conflict with the concept of instantaneous "action at a distance" supported by Helmholtz, but is in accord with Maxwell's theories of electromagnetic wave propagation.

Hertz was honored by the Royal Society, London, which presented him with the Rumford Medal. He is reported as "being of an amiable disposition, genial, a good lecturer, possessed of singular modesty who . . . even when speaking of his own discoveries, never mentioned himself." He died in 1894, at the age of 37.

He can have had no premonition of the ultimate value of his work, for, when asked whether there was a possibility that the waves that came to be called "Hertzian" could be used for telegraph and telephone communication between countries, he replied, "No, it would need a mirror as large as a continent."

Heinrich Rudolf Hertz can, nevertheless, justly be regarded as the father of radio communication.

JAMES CLERK MAXWELL (1831–1879)

Maxwell's biographer Rollo Appleyard has said, "Throughout the history of natural science there is no name except Newton's more honoured by men of science than that of James Clerk Maxwell."

It was Maxwell's remarkable achievement to have demonstrated, by reasoning and pure mathematics, embodied in Maxwell's equations, the fundamental relationships between electricity, magnetism, and wave propagation that underlie all radio and cable communication. He showed

Figure 2.7. James Clerk Maxwell and his electromagnetic field equations.
Maxwell's equations of the electromagnetic field, expressed in mks units, are:

$$\operatorname{curl} E = - \dot{B}, \qquad \operatorname{div} B = 0$$
$$\operatorname{curl} H = \dot{D} + J, \qquad \operatorname{div} D = \rho$$

where

$$B = \mu H, \qquad D = \varepsilon E$$
$$\mu = \mu_o K_m, \qquad \varepsilon = \varepsilon_o K_e$$

K_m and K_e are respectively the magnetic and electric specific inductive capacitances, and μ_o and ε_o the magnetic and electric inductive capacitances of a vacuum; $J =$ current density and ρ = charge density.

that light, like radio waves, is an electromagnetic phenomenon and that both have a maximum velocity of 300 million meters per second. His work provided a theoretical basis for believing that radio waves, like light, could be focused by conducting reflectors—as Hertz later demonstrated practically. Maxwell's equations remain a fundamental starting point for much electrical design, not only for cables both electric and optical but also for the microwave components and aerials used in present-day radio-relay, satellite communication, and radar systems.

Maxwell also had an interest in color vision, shown by the experiments he conducted to demonstrate the nature of light, and which later had a bearing on the development of color television.

He was born in Edinburgh in 1831; he attended Edinburgh Academy where his talent won him prizes in English and mathematics. At 16 he entered the University of Edinburgh and his mathematical abilities flourished under the tuition of Sir William Hamilton, inventor of the mathematical concept "quaternions." In 1850 he went to Cambridge University, where he came under the influence of Sir George Stokes, mathematician and physicist. In 1854 he gained a fellowship at Trinity College and besides lecturing took an interest in Faraday's work on electromagnetism. He became for a time a professor at King's College, London where he developed a close friendship with Faraday.

The period 1850–1865 saw many advances in submarine cable telegraphy, bringing both data and problems for Maxwell to solve.

During a period of illness and temporary retirement to his native Scotland, he began his great work *Electricity and Magnetism,* completed in 1873. In 1871 he was given the Chair of Experimental Physics at Cambridge University, a move that later involved him in the creation of the world-famous Cavendish Laboratory.

The quality of Maxwell's thinking, and his outstanding ability to crystallize subtle electromagnetic phenomena into meaningful quantitative terms can be exemplified by the following statement of two of his four theorems:

1. If a closed curve be drawn embracing an electric current, then the integral of the magnetic intensity taken round the closed curve is equal to the current multiplied by 4π.
2. If a conducting circuit embraces a number of lines of magnetic force, and if for any cause whatever the number of these lines is diminished, an electromotive force will act round the circuit, the total amount of which will be equal to the decrement of the number of lines of magnetic force in unit time.

Maxwell's theory owes something to Ohm's law, and a great deal to Faraday, who saw "lines of force" where other mathematicians saw "action at a distance."

When he applied his equations to electromagnetic wave propagation Maxwell came to the conclusion that

The velocity is so nearly that of light that it seems we have strong reason to conclude that light itself—including radiant heat, and other radiations if any—is an electro-magnetic disturbance in the form of waves propagated through the electro-magnetic field according to electro-magnetic laws.

It is reported that Maxwell was "genial and patient, had great power of concentration . . . considerable knowledge and discrimination in literature . . . was a rapid reader and had a retentive memory . . . he loved his friends, his dog and his horse."

He died in 1879, at the age of 48. To those proceeding to extend his victories along the road of electrical communication, he left his "sword and chariot"—his equations and his theories. He also left an example of individual thought and achievement and a plea for fellowship between all men of science.

WILLIAM THOMSON (LORD KELVIN) (1824–1907)

William Thomson was knighted when 34 years old for his part in the laying of the first transatlantic telegraph cable (Chapter 3). He was given a peerage by Queen Victoria in 1896, becoming Lord Kelvin, and was regarded in his day as the foremost physicist and electrical engineer in the world.[4]

During his long, happy, and vastly productive life, Thomson contributed notably to thermodynamics, electricity and magnetism, electrical engineering, telegraphy, hydrodynamics, navigation, and mathematics.

He was born in Belfast, Northern Ireland, in 1824. His father was Professor of Mathematics at Glasgow University where his son William later became Professor of Natural Philosophy for almost the whole of his working life. William matriculated at the remarkably early age of 10, attending his father's lectures and showing considerable aptitude for languages, logic, and mathematics.

He entered Cambridge University (Peterhouse) at the age of 17, specializing in mathematics. William took an active part in university life, being a good oarsman and a performer in the Cambridge Musical Society. Like many another student he had difficulty in making cash ends meet and his father—a good Scot—frequently wrote to adjure him to "be more careful with his money"!

After graduating from Cambridge and studying in the Regnault laboratories in Paris for a time, Wiliam was awarded the Chair of Natural

Figure 2.8. Lord Kelvin. The "cable" galvanometer used to receive the first transatlantic signals in 1858.

Philosophy at Glasgow University at the remarkably early age of 22, where he remained as Professor of Physics for 53 years. In spite of this long tenure of this post, his versatility and creativity remained at an astonishingly high level throughout.

Very early in his career Thomson became interested in the problems of heat flow and his first paper on this was published in the *Proceedings of the Cambridge Mathematical Society* when he was 15. His work on heat as a form of energy was both creative and profound: it led to the establishment of an absolute zero of temperature at $-273°C$ as a datum for the scale of temperature later designated, in his honor, as "degrees Kelvin." It also contributed to the formulation of the keystone first and second laws of thermodynamics.

The first law demonstrated that, taking into account gains and losses of energy, including heat, in a closed energy system there was at all times a balance of energy in the system.

The second law proved that heat can only change into useful work in flowing from a higher temperature T_1 to a lower temperature T_2, and not in the reverse direction. Sometimes called "time's arrow," this law has significant philosophical implications. It also enabled the formula

$(T_1 - T_2)/T_1$ to be derived for the optimum efficiency of heat engines that set a target for improving the often grossly inefficient steam and other engines.

Thomson's interest in the relation between heat and electricity led him to study the Seebeck effect that relates to the generation of electricity by heat applied at a junction between a pair of dissimilar conductors, and the Peltier effect that is concerned with the cooling at such a junction when carrying an electric current. His work showed the quantitative relationship between these two effects and revealed a third effect—Thomson heat. These results were to prove of major importance in the design of refrigeration systems.

His electrical inventions were numerous and valuable; they included an electrostatic voltmeter, a quadrant electrometer, and a balance for the measurement of electric currents. But perhaps the most useful from a telecommunication point of view was the highly sensitive mirror galvanometer that played such an important role in the first transatlantic telegraph cable (see Chapter 3). He invented means for correcting the magnetic compass on board ships from errors related to the ship's magnetic field and improved the mounting of the compass to minimize the effect of rolling—further assisting in overcoming the difficulties of submarine cable laying.

Sir J. J. Thomson (no relation to William Thomson), then a professor at the Cavendish Laboratory, Cambridge, said of Kelvin:

> Modern wireless telegraphy, telephony and broadcasting depend on a result published by him (Kelvin) in 1853.

The result referred to appears in a paper by William Thomson on "Transient Electric Currents" in the *Philosophical Magazine*, June, 1853, in which he derived solutions for the behavior of current in a damped oscillatory circuit, including the result that, when the damping is light, the resonant frequency is given by

$$f = 1/2\,\pi\,\sqrt{LC}$$

where L and C are the inductance and capacitance of the oscillatory circuit. This equation is written in the mind, if not on the heart, of all good radio engineers today!

Kelvin's thinking was not, however, limited to the specific scientific problems of everyday life; his second law of thermodynamics, with its implication of "time's arrow," led him to speculate on the age of the Earth and indeed the universe itself. Although his estimate of the age of the Earth was in error by a substantial margin, he nevertheless came to the conclusion that all things had a beginning (supported by the present-day "big bang" theory) and that "if a probable solution, consistent with ordinary course of nature, can be found, we must not invoke an abnormal act of Creative Power."

Kelvin died in 1907, with many honors including fellowships of the Royal Societies of Edinburgh and London (granted when he was only 22 and 26 years of age, respectively), and was buried in Westminster Abbey, fittingly near the grave of Sir Isaac Newton.

He had a remarkable simplicity of character and a full measure of scientific modesty. At his Professorial Jubilee in Glasgow University in 1896, after he had played such a great role in the development of telegraphy and electrical engineering, he said rather wistfully:

> I know no more of electric and magnetic force . . . than I knew and tried to teach my students of Natural Philosophy fifty years ago.

THE ELECTRON IS DISCOVERED AND THE ELECTRONIC AGE BEGINS

Once the laws of electromagnetism and the identity of light with electromagnetic waves had been established, the next major advance was the discovery of the electron—a discovery that led to an understanding of the nature of atomic structure and the development of quantum physics vital to the creation of new electronic devices such as the thermionic valve, the transistor and the microchip, the cathode-ray picture tube, lasers as sources of coherent light, and light-sensitive photodiodes. The first step——the discovery of the electron—was made by Sir J. J. Thomson.

Sir J. J. Thomson (1856–1940)

J. J. Thomson, the son of a Manchester bookseller, entered a local college at the age of 14 for a career in engineering. However, he found experimental physics more interesting and in 1876 won a scholarship to Cambridge University where he remained for the rest of his life. In addition to his outstanding abilities in mathematics and physics, he was a skilled administrator and his advance at Cambridge was spectacular. At the age of 27 he succeeded the distinguished physicist John Rayleigh as

Figure 2.9. Pioneers of the electronic age: Sir. J. J. Thomson (top) and Sir Ernest Rutherford. (a) A discharge tube designed by Thomson in which cathode rays were allowed to travel past electrically charged plates. The deflection of the cathode rays by electric and magnetic fields convinced Thomson that these rays consisted of particles, which we now know as electrons. (b) Rutherford's first notes on the structure of the atom deduce equations relating the negative charge on the electron and the radius of its orbit to the postive charge on the nucleus.

Professor of Physics and became responsible for the Cavendish Laboratory, which acquired worldwide fame for research in atomic physics under his direction.[5]

Thomson's discovery of the electron arose from his interest in cathode rays—the radiation from the negative electrode or cathode when an electric current passed between two electrodes in an evacuated glass tube. By measuring the deflections of the cathode rays in the presence of electric and magnetic fields of known strengths, Thomson was able to show that the cathode rays were streams of negatively charged particles he named *corpuscles,* later called *electrons*. From the deflection of the cathode rays by the electric and magnetic fields he deduced the ratio of the electric charge to the mass of the individual electrons, and showed that these were much smaller in size than the atoms themselves.

Support for Thomson's discovery came a few years later in 1886 when the German physicist Eugen Goldstein showed that the electron stream in a cathode-ray tube was accompanied by a stream of positive charges—he called them *canal rays*—moving in the opposite direction. These were ions, most loosely bound atoms stripped of their electrons. He further demonstrated that if the cathode had a hole, the canal rays produced a luminous spot on a screen behind the cathode.

Although the modern cathode-ray tube uses an electron stream rather than positively charged particles to produce waveform and picture displays, its origin can clearly be seen in the ideas initiated by Thomson and Goldstein in the late 19th century.

Thomson speculated about the structure of the atom in terms of charged particles, the negative ones being akin to his "corpuscles." This theme was later followed up by Rutherford (see below).

Thomson was one of the great intellectual celebrities of British science. He was also an inspired teacher and he gathered around him at the Cavendish Laboratory some of the most brilliant young minds of the day, many of whom became professors in the world's leading universities and seven of whom won Nobel Prizes. One of Thomson's former students was Sir Ernest Rutherford who later achieved much fame in the field of atomic physics.

Sir Ernest Rutherford (1871–1937)

Ernest Rutherford was born in New Zealand where his father was a farmer. Ernest in his younger days shared the hard work of running the farm. He showed great promise as a student and won a scholarship at Canterbury College in Christchurch.

He graduated with honors in 1882 and won another scholarship which took him to Cambridge University. There he met Sir J. J. Thomson who encouraged him to work on X rays, which had just been discovered by the German physicist Wilhelm Roentgen. For Rutherford this was the beginning of a lifetime interest in radioactivity and atomic structure. The former led to many studies of both the properties of the emissions—alpha, beta, and gamma rays—from the radioactive elements in the uranium and thorium series which terminate in stable lead, and the exponential decay of the rate of emission from any active sample expressed in terms of the "half-life," at which the rate has been reduced to half.

In a classic series of experiments in 1909 he analyzed the pattern of particles coming from a gold target bombarded by a beam of alpha particles, giving him an insight into the structure of atoms. He formulated a model in which the mass of the atom was concentrated in a positively charged nucleus, around which electrons moved much as do planets around the sun. This model was refined, notably by Niels Bohr, and provided a foundation for the quantum theory,[6] which enabled a wide range of apparently unconnected physical phenomena to be explained.

His work shows the remarkable power of an original mind to find commonsense answers to complex scientific problems. His was a warm and humorous nature that won him many friends. His strong opposition to the rise of the Nazi party in Germany during the pre-War years led him to help many Jewish refugees to flee from the Nazi persecution. In 1931 he was honored with the title of Baron Rutherford of Nelson. With J. J. Thomson he helped to lay the foundation of modern electronics.

Lord Rayleigh (1842–1919)

The scientific career of John William Strutt, 3rd Baron Rayleigh, covered a period when physical science was undergoing remarkable development at the hands of his contemporaries such as James Clerk Maxwell, Oliver Heaviside, and Lord Kelvin. A man of vast erudition, broad interests, and tremendous energy, he made important contributions to every branch of physics known in his day, including heat, light and sound, electricity and magnetism, and the properties of matter.[7]

Rayleigh's *Scientific Papers* (Dover Publications, New York, 1964) comprise over 400 writings; perhaps the more important for their later impact on the development of telecommunications and broadcasting were the following:

- Rayleigh's "Theory of Sound," one of the great classics in the literature of physics
- His work on "The Scattering of Light by Small Particles," which laid a foundation for later theories of radio wave propagation by scattering in the troposphere and ionosphere
- The determination of electrical standards, notably the ohm
- "The Passage of Electric Waves Through Tubes, or the Vibrations of Dielectric Cylinders," which became a starting point for the theory of long-distance waveguide and optical fiber transmission systems (Chapters 16 and 17).

Lord Rayleigh was born John William Strutt, eldest son of the 2nd Baron Rayleigh of Terling Place, Witham, in the county of Essex, England. From his earliest schooling days he showed an aptitude for mathematics and in 1861, when nearly 20, he entered Trinity College, Cambridge. Here his mathematical bent was stimulated by the famous mathematical "coach" E. J. Routh and Sir George Stokes, the Lucassian Professor of Mathematics who was also greatly interested in experimental physics. In the Cambridge Mathematical Tripos of 1865, Rayleigh emerged as Senior Wrangler and in 1886 was elected as Fellow of Trinity College.

The establishment of the Cavendish Laboratory at Cambridge with James Clerk Maxwell as the first Professor of Experimental Physics provided a further stimulating background to Rayleigh's work. On Maxwell's untimely death in 1879, Rayleigh succeeded to the Professorship and initiated a comprehensive teaching program in heat, electricity and magnetism, optics and acoustics, and the properties of matter— a pioneering effort that had a major impact on physics teaching in the United Kingdom and Europe. However, much of his experimental work was carried out in his rather primitively equipped laboratory at his home in Terling Place, Essex.

He became President of the Mathematics and Physics Section of the British Association and Secretary, later President, of the Royal Society, London. In 1887 he was appointed as Professor of Natural Philosophy at the Royal Institution of Great Britain, where Faraday had conducted his epoch-making investigations into electromagnetism.

It was at the Royal Institution that Rayleigh began the famous "Friday evening" discourses and the Christmas "Children's Lectures" that are so popular today as a means for promoting a continuing interest in science.

Public recognition came to Lord Rayleigh in full measure; in 1904 he received the Nobel Prize for his discovery of argon, and he was one of the first recipients of the Order of Merit in 1902. Many honorary degrees and awards from learned societies followed. His contemporaries and successors alike place Rayleigh in that great group of 19th-century physicists that made British science famous throughout the world. He was above all a modest man, with a great sense of humor. When he received the Order of Merit from the Crown he is reported to have said that "the only merit of which he personally was conscious was that of having pleased himself with his studies, and any results that may have been due to his researches were owing to the fact that it had been a pleasure to become a physicist."

Max Planck (1858–1947)

The German physicist Max Planck was foremost in demonstrating that energy, at the atomic level, exists in indivisible packets or quanta. In the case of light the unit is the photon, the energy of which is directly proportional to the light frequency according to the law

$$\text{Photon energy } E = \text{Constant } h \times \text{Frequency } f$$

The constant h is known as "Planck's constant." This law, and the concept of energy quanta, have had a profound influence on the development of atomic physics. It is at the heart of the technological design of devices such as television camera and display tubes, lasers, and light amplifiers for optical telecommunication systems.

Max Planck was born in Kiel, Germany; he studied at Munich University and later in Berlin. His early work was in thermodynamics and on "blackbody radiation," i.e., a body that radiates perfectly at all frequencies, work that led him to the principle of energy quanta. For his work in the quantum field he was awarded the Nobel Prize in 1918. His collaboration with Albert Einstein before and after World War I made Berlin a world center for the study of theoretical physics.

REFERENCES

1. Rollo Appleyard, *Pioneers of Electrical Communication* (Macmillan & Co., London, 1930).
2. *Pioneers in Telecommunications* (British Telecom Education Service, BT HQ, 81 Newgate St., London, 1985).
3. William Cramp, *Michael Faraday* (Sir Isaac Pitman and Sons Ltd., London, 1931); John Meurig Thomas, *Michael Faraday and The Royal Institution* (Adam Hilger, Bristol, 1991).
4. D. K. C. Macdonald, *Faraday, Maxwell and Kelvin* (Heinemann, London, 1965).
5. A. Feldman and P. Ford, *Scientists and Inventors* (Aldus Books, London, 1979).
6. T. Hey and P. Walters, *The Quantum Universe* (Cambridge University Press, London, 1987).
7. R. B. Lindsay, *Lord Rayleigh: The Man and His Work* (Pergamon Press, Elmsford, NY, 1970).

Further Reading

J. G. O'Hara and W. Pricha, *Hertz and the Maxwellians*, IEE History of Technology Series, No. 8 (London, 1987).
P. Tunbridge, *Lord Kelvin: His Influence on Electrical Measurements and Units*, IEE History of Technology Series, No. 18 (London, 1992).
G. R. M. Garratt, *The Early History of Radio: From Faraday to Marconi*, IEE History of Technology Series, No. 20 (London, 1994).

3

The First Telegraph
and Cable Engineers*

THE VISUAL TELEGRAPH (SEMAPHORE)

Claude Chappe (1763–1805)

The first practical system of long-distance telegraphic communication was the semaphore, which used manually movable arms or closable apertures mounted in towers, usually on hilltops and within line-of-sight of one another, to convey messages letter by letter.

The French Revolution stimulated the need for swift and reliable communication throughout France in the 1770s—a need met by the brilliant French engineer Claude Chappe. Eventually 500 semaphore stations, spanning some 5000 km, had been installed throughout France.

News of Chappe's invention reached the British Admiralty and stimulated Lord George Murray (1761–1803) to develop a visual tele-

---

*In this chapter use has been made of material originally published by British Telecom in Ref. 1, the International Telecommunication Union in Ref. 2, and the Science Museum, London in Ref. 3.

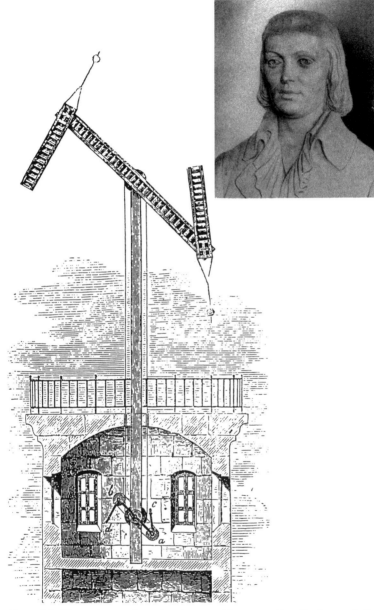

Figure 3.1. Chappe's semaphore showing movable arms and handles to operate them. (Top right) Bust of Claude Chappe.

graph system using holes closed manually by movable wooden shutters. A system of 15 stations was installed between London, Deal, and Portsmouth—some of them on sites still known as "Telegraph Hills."

The semaphore may be regarded as a precursor of present-day microwave radio-relay systems which often use hilltop sites. However, the semaphore with its dependence on the transmission of light through the atmosphere was susceptible to severe disruption in bad weather. And the rate of transmission of information was measured in words (or tens of bits) per minute, compared with 100 million bits per second or more on microwave systems.

THE ELECTRIC TELEGRAPH

The discoveries in electromagnetism in the early 19th century—notably Volta's battery, Faraday's and Oersted's discoveries on the relation between electricity and magnetism—created the possibility of the electric telegraph, one using the transmission of an electric current along a conducting wire. Unlike the semaphore, this device clearly was immune to the effects of weather and offered speedier transmission of information.

Although there had been attempts to devise an electric telegraph from 1800 onwards in Britain, Russia, Bavaria, and Switzerland, none of these found practical use on any scale.

It was the partnership of Wheatstone and Cooke in the 1830s that led to the first electric telegraphs in the United Kingdom, inspired mainly by the need of the expanding railway system for a fast and reliable means of communication to ensure regular running of the trains, and especially standardized timekeeping throughout the country.

Charles Wheatstone (1802–1875) and William Fothergill Cooke (1806–1879)

Charles Wheatstone was born in 1802 in Gloucester into a family of musical instrument makers; his interest in music led him to experiment with sound and how it was produced. Other experiments included an attempt to measure the speed of propagation of an electrical current along a wire. For his experimental work he was elected a Fellow of the Royal Society in 1836.

Figure 3.2. Charles Wheatstone (left) and William F. Cooke.

W. F. Cooke was born in 1806; his father was a surgeon and a professor of anatomy. William's early life gave little indication of future scientific interests. After leaving university he became a soldier in the East India Company army, but retired after a few years on health grounds.

In 1836 Cooke attended some telegraphic experiments that Professor Munke at Heidelberg was making with a needle galvanometer, and was seized with the idea of turning this into a commercially practicable telegraph system. He consulted Michael Faraday, and then Professor Wheatstone at King's College, London, who had also been conducting experiments with telegraphic apparatus. He built his own apparatus and wrote a pamphlet about it, suggesting that telegraph wires could easily run alongside railway tracks.

When Cooke and Wheatstone met in 1837, both were convinced of the possibilities of the telegraph and decided to form a partnership to develop and exploit it, taking out their first patent on July 10, 1837. This described a telegraph system using five magnetic needles, each of which could be deflected by an electric current as in Oersted's original experiment. By deflecting the needles successively in pairs, any one of 20 letters of the alphabet could be selected in turn. The deflection of one needle could be used to point to a numeral. This system initially required five wire lines, later reduced to two by using a code.

The directors of the Great Western Railway were impressed by the invention and some 20 km of line was installed in 1838 between Paddington and West Drayton, using the five-needle telegraph.

By 1852, 4000 miles of telegraph was in use in the United Kingdom. Cooke and Wheatstone continued to improve their telegraph system, eventually using only one needle and a signal code. There was, however, considerable ill-feeling between the two men as to who was the sole inventor of the electric telegraph. Perhaps the most appropriate comment is that both men were eventually knighted for their achievements in telegraphy.

Developments in electric telegraphy were also taking place in the United States, notably by Morse.

Samuel F. B. Morse (1791–1872)

Samuel Morse was born in 1791 at Charlestown, Massachusetts; he studied at Yale College, graduating in 1810. He had a talent for painting and studied in Europe, having a picture exhibited in the Royal Academy, London. He returned to the United States in 1815, but his hopes for a career in painting were only partially fulfilled by portraiture. In 1825 his wife and parents died and he had lost money in failed artistic ventures and paying debts incurred by his father.

However, he had learned about electricity at college and it had become a hobby for him in the 1820s. He attended lectures on electro-magnetism in New York in 1826 and, in 1832, met a fellow passenger on a return trip from Europe, with whom he discussed and made notes on the possibilities of using an electromagnet as the receiving element in a telegraph system.

By 1835 he had made a "printing telegraph" in which a key could switch on the current to an electromagnet as long as the key was held down, so causing a pencil to make long or short marks on a moving strip of paper.

Morse's key invention was undoubtedly the code that bears his name—the Morse code. He assigned different combinations of dots and dashes to the letters of the alphabet and the numerals. Experienced operators learned to "read" the code by listening to the sounds of the clicking electromagnet, without having to see the dots and dashes on the paper tape, thus giving rise to the simplest of all telegraph systems using a key, battery, line, and sounder. The British "needle" telegraph

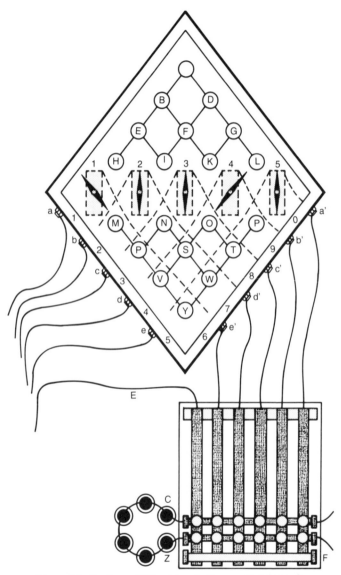

Figure 3.3. Cooke and Wheatstone's five-needle telegraph.

Figure 3.4. Samuel F. B. Morse.

was adapted to use the Morse code by interpreting a right deflection as a "dot" and a left as a "dash."

Morse had a difficult time launching his invention and he nearly starved until, in 1843, Congress finally gave him funds to set up a trial system between Baltimore and Washington in 1844. From that time forward, telegraph systems expanded throughout the world and Morse, after several lawsuits, became a wealthy man.

It is difficult to overestimate the value and importance of the Morse code. Not only did it provide the basis for a simple, robust, and easily operated overland line telegraph system, it also served for point-to-point and mobile (e.g., ship-to-ship and ship-to-shore) radio communication. For such applications the ability of the Morse code to be heard and interpreted by a skilled operator under conditions of high levels of radio interference and noise is of paramount importance, especially for distress calls. These are, no doubt, reasons why the Morse code continues in widespread use today.

But perhaps of even greater importance is the fact that the concept—inherent in the Morse code—of transmission of information by coded groups of on/off signals has led to a major revolution in commu-

E	–	12,000
T	—	9,000
A	– —	8,000
I	– –	8,000
N	— –	8,000
O	– –	8,000
S	– – –	8,000
H	– – – –	6,400
R	– – –	6,200
D	— – –	4,400
L	—	4,000
U	– – —	3,400
C	– – –	3,000
M	— —	3,000
F	– — –	2,300
W	– — —	2,000
Y	– – – –	2,000
G	— — –	1,700
P	– – – – –	1,700
B	— – – –	1,600
V	– – – —	1,200
K	— – —	800
Q	– – — –	500
J	– – – –	400
X	– — – –	400
Z	– – – –	200

Figure 3.5. Morse's original code showing relation of lengths of dot and dash groups to the quantities of type found in a printer's office.

nications: the use of digital transmission and switching for all types of information, including audio, data, facsimile, and television.

By Morse's death in 1872, other inventors and engineers had taken over from him, improving his system—Edison in the United States, Werner Siemens in Germany, and his brother William in England.

THE PRINTING TELEGRAPH

David Edward Hughes (1831–1900)

David Hughes was born in London in 1831. His family emigrated to the United States in 1838, and in 1850 he became Professor of Music at

St. Joseph's College, Bardstown, Kentucky. Five years later he invented a letter-printing telegraph with a 52-symbol keyboard in which each key caused the corresponding letter to be printed at the distant receiver. This was achieved by a rotating wheel bearing 28 letters of the alphabet and other signs on it, and a clutch mechanism actuated by an electromagnet which brought the wheel momentarily to rest when the desired letter was above a moving strip of paper—an arrangement with some similarities to the familiar type "golf ball" in a modern typewriter. When the sending operator pressed one of the keys on the keyboard, the electric impulse sent to line stopped the rotating type-wheel at just the right instant of time to cause the desired letter to be printed.

Hughes's achievement was all the greater because the typewriter had not yet been invented! The modern teleprinter, the "telex" system, and computer keyboards are direct descendents of Hughes's invention and are now the standard means of telegraphic communication in postal, newspaper, and business services.

DUPLEX, QUADRUPLEX, AND TIME-DIVISION MULTIPLEX TELEGRAPHY

As telegraph traffic grew, inventors sought means whereby each cable or wire could carry two or more messages simultaneously, thereby saving the cost of extra cables or wires—an important consideration on long-distance routes.

The first to use a single telegraph wire for sending two messages simultaneously, one in each direction, was Dr. Gintl in Vienna in 1853, His "duplex" circuit involved the use of an artificial line simulating the real line in a balanced bridge arrangement, thereby enabling a message to be sent at the same time that one was being received, without mutual interference.

Thomas Alva Edison (1847–1931), who had worked as a telegraph operator from the age of 15, turned his brilliantly inventive mind to the improvement of the telegraph. His patents included a duplex circuit and, in 1874, a quadruplex circuit enabling the simultaneous transmission of four messages on a single wire.

The next important development in telegraphy was the invention of time-division multiplexing—attributed to Baudot.

Emile Baudot

Emile Baudot was an officer in the French Telegraph Service who invented a system based on the use of a five-unit code, each letter of the alphabet being represented by a unique combination of the five elements. He combined the use of the five-unit code with time-division multiplexing—the latter being achieved by a rotating commutator at the receiving end of the line, synchronized with a similar commutator at the sending end. This enabled a single wire line to be used for the simultaneous transmission of as many messages as there were segments on each commutator. Each sending operator had a keyboard with five keys, corresponding to the five-unit code, while at the receiving end the message appeared on a strip of paper passing through a printer.

Baudot's system was first introduced in 1874 and officially adopted by the French Telegraph Service in 1877.

His invention is particularly significant because it embodied two principles that were later to become of even greater importance: (1) the representation, as with Morse, of a symbol by a code and (2) the concept of time-division multiplexing. The first foreshadowed pulse-code modulation, while the second was a forerunner of modern digital techniques.

THE FIRST UNDERSEA TELEGRAPH CABLES

Once overland telegraph cables had been established in Great Britain and the continental countries, there was a clear need—for social, news, business, and diplomatic purposes—to provide links across the English Channel.

The first Dover–Calais telegraph cable was laid in 1850 but proved vulnerable to the operations of trawlers. A second cable, with four copper conductors insulated with gutta-percha and protected by tarred hemp and armoring wires, was laid in the following year and proved more successful, remaining in operation for many years.

This success stimulated the provision of telegraph cables across the North Sea and in other parts of the world, e.g., Italy to Corsica and Sardinia, India to Ceylon, and Tasmania to Australia.

However, although land links to British possessions in the Middle and Far East were possible, these were vulnerable to both technical breakdown and political disruption or interception in times of trouble. Thus, for strategic as well as practical considerations, an undersea telegraph cable system from the United Kingdom via Gibraltar, the Mediterranean, the Red Sea to Singapore, China, and Australia was projected.

The initiative taken by Great Britain in setting up a worldwide telegraphic network prompted Reuter to set up his headquarters in London, which became established as the world's first telegraphic news and financial information distribution center.

There remained, however, the most important cable yet to be laid, the one to span the Atlantic Ocean.

THE FIRST TRANSATLANTIC TELEGRAPH CABLE

Cyrus W. Field (1819–1892) and William Thomson (1824–1907)

The story of the first transatlantic telegraph cable is an epic of courage, enterprise, and perseverance. It owes much to a great American citizen, Cyrus W. Field, whose untiring efforts provided a major impetus throughout the project. But equally important was the scientific advice given by Professor William Thomson (later Lord Kelvin).

The first attempts in 1857 to lay the cable between Valentia, Ireland, and Newfoundland by British and American warships were unsuccessful because of breaks in the cable caused partly by unsatisfactory cable-laying techniques.

A repeat attempt in 1858 was more successful and on August 14, 1858, Queen Victoria and U.S. President James Buchanan exchanged messages of congratulation. But within a month of the first messages the cable failed: an insulation breakdown was caused by an overzealous operator who had applied an excessive voltage in an effort to improve the signals.

The U.S. Civil War delayed work on the cable for a time, but by 1865 the Atlantic Telegraph Company had raised new capital, redesigned the cable, and chartered Brunell's *Great Eastern* with improved laying equipment. The *Great Eastern* was a huge passenger liner, unsuccessful as such, but was the only ship afloat with capacity to hold the entire 3700 km of cable.

Professor William Thomson (see Chapter 2) had been engaged as a consultant to advise and report on the electrical performance of the

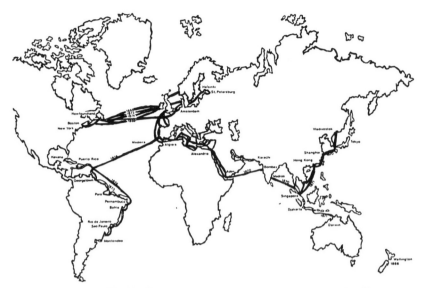

Figure 3.6. Worldwide network of submarine telegraph cables (1875).

cable during the design stage and on board the *Great Eastern* during the laying operations.

A long cable functions as large distributed capacitance and this severely limits the rate at which the signal current waveform can rise or fall, and therefore restricts the rate at which messages can be sent over the cable. This in turn decides the earning capacity of the cable and the ultimate financial return to the cable company. The theoretical prediction of the electrical performance owes much to the work of Kelvin and Heaviside, as does the design of the sending and receiving equipment for optimum performance. Kelvin's invention of the highly sensitive mirror galvanometer was of critical importance in the operation of the cable.

The first of the renewed attempts in 1865 to lay the cable was a failure caused by a breakage and the *Great Eastern* had to return to Valentia. A second attempt in 1866 was successful and on July 27 the first telegraph messages were transmitted between the United Kingdom and the United States.

Not only was the first successful transatlantic telegraph cable of great importance as the only communication channel then available between the two countries, it also gave valuable practical experience in overcoming the difficult problems of laying and recovering cables in the great depths of the Atlantic Ocean. This experience, and the confidence it generated, were of immense value when, nearly a century later, the first transatlantic telephone cable came to be laid.

By 1875 a worldwide network of telegraph cables had been laid, linking the United Kingdom, the United States, India, the Far East, and Australia—a tribute to the telegraph pioneers of Victorian England and the United States.

REFERENCES

1. *Pioneers in Telecommunications* (British Telecom Education Service, BT HQ, 81 Newgate St., London, 1985).
2. *From Semaphore to Satellite* (The International Telecommunication Union, Geneva, 1965).
3. Bernard S. Finn, *Submarine Telegraphy: The Grand Victorian Strategy* (Science Museum, London, 1973).

4

The First Telephone Engineers

THE TELEGRAPH LEARNS TO TALK:
THE INVENTION OF THE TELEPHONE

It may well have been Alexander Graham Bell's successful demonstration on March 10, 1876, of what was later claimed to be the world's first demonstration of the electronic transmission of intelligible speech—enshrined in Bell's historic call to his assistant: "Mr. Watson, come here, I want to see you"—that set in motion what eventually became a major step forward in communication at distance. However, the history of the invention of the telephone is complex and Bell's was by no means the only contribution.

The need for the telephone, as compared with the telegraph, is clear. The world of the 19th century was that of the Industrial Revolution; science, technology, manufacture, and trade were on the march, and new political forces were making themselves felt. There was a growing need to communicate, rapidly and in the most natural way possible. The electric telegraph met these requirements only partially, being slow and conveying little of the personality of the communicators. Speech, developed during the many millennia of human evolution and closely linked to the maximum rate at which we could create, absorb, and respond to information, was clearly the most natural and useful mode of person-to-person communication.

Fortunately the discoveries in the field of electromagnetism made by Oersted, Faraday, Henry, and others in the early 19th century had not only made possible the electric telegraph, they also created the foundation of scientific knowledge necessary for the invention of the telephone. Furthermore, by 1875 there was already in existence an extensive and growing network of intercity and international telegraph wires and cables that must have prompted the question: if these could carry the electric currents that corresponded to telegraph signals, might they not be adapted to carry electric signals corresponding to speech?

Alexander Graham Bell (1847–1922)

Alexander Graham Bell was born in Edinburgh in 1847. Both his father and grandfather were interested in the nature of speech and the human vocal system, and this no doubt stimulated a similar interest in Alexander. The family moved to Canada and later to Boston where Bell became Professor of Vocal Physiology at Boston University.

He was a successful teacher of deaf children, making a notable contribution by disproving an earlier contention that the deaf could not

Figure 4.1. Alexander Graham Bell (left) and Thomas Alva Edison.

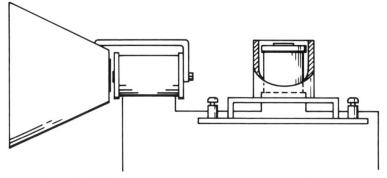

Figure 4.2. Bell's electromagnetic telephone transmitter and receiver.

be taught to speak.[1] Besides teaching the deaf and dumb, Bell was absorbed in the scientific study of sound and the possibilities of transmitting sound by electricity or light.

Bell's telephone comprised a transmitter and a receiver each with a thin iron diaphragm in front of an iron core surrounded by a coil of wire, the iron core being magnetized, e.g., by direct current in the coil from a battery. Sound waves falling on the transmitter diaphragm caused it to vibrate and generate a similar variation of the magnetic field. This in turn induced an undulating electric potential and current in the coil—a direct application of Faraday's principle of electromagnetic induction. The undulating current from the transmitter was transmitted over conducting wires to the coil in the distant receiver where it generated a varying magnetic attraction on the diaphragm, thus producing a copy of the original sound waves. This is the arrangement described in Bell's first patent, No. 174,465 of March, 1876. His second patent, No. 186,787 of January, 1877, described the use of a permanent magnet, enabling the battery to be dispensed with. This patent included, almost as an afterthought, and coincident to a caveat entered by fellow inventor Elisha Gray, a reference to a liquid variable-resistance transmitter.

It is remarkable that modern telephone receivers use principles identical to those established by Bell in 1876. However, a marked improvement in the sensitivity of the transmitter was achieved, perhaps at some expense of sound quality and long-term reliability, by Thomas Alva Edison's invention of the carbon granule transmitter, the resistance of which varied in sympathy with the sound wave pressure on it. This,

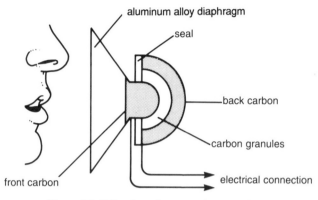

aluminum alloy diaphragm

seal

back carbon

carbon granules

front carbon

electrical connection

Figure 4.3. Edison's carbon granule transmitter.

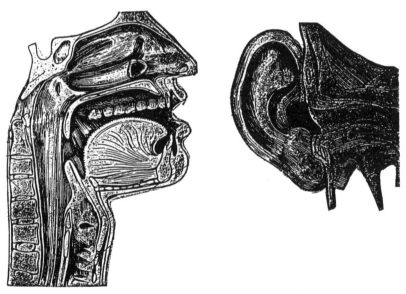

Figure 4.4. The world's most advanced form of communication—human speech.

until recently, was used in virtually all telephone systems. Edison's invention of the induction coil—two coils of wire on an iron core, much as in one of Faraday's experiments—further improved the telephone by enabling the battery circuit to be localized and by providing a better impedance match to the line.

Bell's telephone was exhibited at the United States 100th Anniversary of Independence Exhibition in Philadelphia in 1876, where the Emperor of Brazil, astounded to hear Bell's voice, exclaimed, "It talks!" It was also demonstrated to Queen Victoria at Osborne, England, in 1877, who is said to have been "most impressed." The Telephone Age had begun.

Figure 4.5. Alexander Graham Bell, inventor of the telephone.

Bell seems to have been modest about his achievement and is on record as saying:

> if he had known more about electricity he would not have invented the principle of the telephone.

However, his words on March 10, 1876, were nevertheless prophetic and showed he had a fair vision of what the telephone might do:

> This is a great day with me and I feel I have at last struck the solution of a great problem—and the day is coming when telephone wires will be laid on to houses, just like water or gas, and friends converse without leaving home.

He foresaw too that

> the telephone would be a new factor in the new urbanization ... without the telephone the 20th century metropolis would have been stunted by congestion and slowed to the primordial pace of messengers and postmen. And the modern industrial age would have been born with cerebral palsy.

THE OTHER INVENTORS OF THE TELEPHONE

The history of the invention of the telephone is complex, confused, and incomplete. Elisha Gray, whose own contribution was by no means insubstantial and who sought to challenge the Bell patents, wrote:

> The history of the telephone will never be fully written. It is partly hidden in 20 or 30 thousand pages of testimony and partly lying in the hearts and consciences of a few whose lips are sealed—some in death and others by a golden clasp whose grip is tighter.

It is clear from the literature that Bell's contribution was by no means the unique and original one that some of his supporters have claimed.[2] Between 1820 and 1860 a score or more of inventors had described, and some had made to work, devices that in part anticipated the essential features of Bell's telephone. These inventors included Meucci (Italy), Bourseul (France), Reis (Germany), Elisha Gray and Thomas Edison (USA).

With the growing commercial value and importance of the telephone, and the massive financial issues that were involved, it is hardly surprising that the 20 years following the issue of Bell's patent saw more than 600 cases of patent litigation, including a long and drawn-out but unsuccessful suit by the U.S. courts against the Bell Telephone Company's monopoly claim to the use of electricity for the transmission of speech.

However, in every case Bell was upheld as the inventor of the telephone, perhaps in no small measure because his demonstration of 1876 clearly revealed its capability for the transmission of good-quality, intelligible, and recognizable speech.

THE SOCIAL AND ECONOMIC VALUE OF THE TELEPHONE

Not everyone in the 19th century was convinced of the value and future prospects of the telephone.

The president of the mighty Western Union Telegraph Company in the United States turned down the opportunity to buy the rights to Bell's patents for a song, thereby inhibiting indefinitely the possibility of a rational and economic integration of the telegraph and telephone services in that country.

Some of the users were not altogether satisfied with the performance of the new device. In 1880 Mark Twain, who was evidently having difficulty hearing over the lines of the Hartford Telephone Company, wrote in a letter to the *New York World* newspaper:

> It is my heart-warm and world-embracing Christmas hope and aspiration for all of us—the high, the low, the poor, the rich, the admired, the despised—may eventually be gathered together in a heaven of everlasting rest and peace and bliss—except the inventor of the telephone!

Others have objected to the telephone for its intrusions into personal privacy; Bell himself refused to have one in his study! Some have criticized the telephone for its adverse effect on letter-writing—a civilized and cultivated art in Victorian times. The speed and immediacy of the telephone are not without disadvantages; for example, the instability and collapse of the Wall Street stock market in 1929 have been attributed in part to the use of the telephone for the panic selling of shares. This has an interesting parallel with the effect of computer-programmed share selling today. And many a military commander in the field has resented "being at the end of a telephone."

But the overall evidence in favor of the telephone is clear and overwhelming since more than 700 million are now in use worldwide.

The value of personal identification provided by the telephone is obvious and far-reaching in many areas of human life, ranging from a young child's first call to a grandparent, to a call for help by a potential suicide, and to a Presidential call on an international "hot line" heading off a nuclear disaster.

As American sociologist Sydney H. Aronson has written,

> it has helped to transform life in cities and on farms, and to change the conduct of business . . . it imparted an impulse towards the development of a "mass culture" and a "mass society", at the same time it affected institution patterns in education and medicine, in law and warfare, in manners and morals, in crime and police work, in the handling of crises and the ordinary routines of life. It markedly affected the gathering of news and the patterns of leisure activity, it changed the context and even the meaning of neighbourhood and of friendship, it gave the family an important means to adapt itself to the demands of modernization and it paved the way both technologically and psychologically for the 20th century media of communication: radio and television broadcasting.[3]

One may be justified in feeling that this was a remarkable outcome from an invention that was in itself of quite remarkable simplicity.

THE BEGINNINGS OF AUTOMATIC TELEPHONE EXCHANGE SWITCHING

The problem of connecting a calling to a called customer was at first solved by manually operated telephone exchanges in which an operator simply plugged in a cord between the corresponding incoming and outgoing telephone line terminals on a switchboard. This system had the advantage that it provided, from the customer viewpoint, good service since the operator could, in systems with small numbers of lines, readily find the called customer by name, and answer queries made by the caller. But it became cumbersome when large numbers of lines were involved, a difficulty only partially solved by the use of "multiple" switchboards with groups of operators. And there was the inherent problem of "overhearing" the customers' telephone conversations by the operators and the consequent lack of privacy.

Almon Brown Strowger (1839–1902)

The lack of privacy was the motivation that in 1889 led a Kansas City undertaker, Almon Brown Strowger, to seek a solution to the interconnection problem by inventing an automatic telephone switching system that dispensed with operators. Strowger's system came to be christened "the girl-less, cuss-less telephone," the latter adjective presumably being justified because it eliminated the delays occasioned by inattentive operators! It has been said that Strowger found that he was losing money in his undertaking business because one of the switchboard operators at the Kansas City telephone exchange was married to a rival undertaker and she would connect Strowger's callers wishing to make funeral arrangements to her husband.

Strowger's invention—which he first modeled with a collar box and matches to represent lines—was of remarkable simplicity. It comprised two basic elements:

1. A device for use by the customer which created trains of on–off pulses of current corresponding to the digits 0–9 (this eventually became the familiar circular 10-hole telephone dial)
2. A switch at the telephone exchange in which a rotating arm was caused to move step by step over a semicircular arc of 10 contacts, each contact being connected to a customer's line, the stepping motion being controlled by pulses of current on the calling customer's line via an electromagnet

In this form, only 10 lines could be accommodated; by providing 10 such arcs of 10 contacts, one above the other, and arranging that the rotating arm stepped first vertically and then horizontally, 100 lines could be accommodated. Eventually step-by-step switches with 100 contacts in each horizontal arc were devised, making possible 1000 line exchanges. Beyond this number of lines, Strowger's original concept became impracticable and uneconomic. New approaches to exchange design were developed in which a Strowger-type switch first found the calling customer's line and then passed the call on to other switches which could be shared with many customers, thereby economizing in the amount of equipment required and making possible 100,000 line exchanges, suitable for large cities.

Strowger's U.S. patent was filed in March, 1889, and issued in May, 1891; the first practical application was at La Porte, Indiana, in 1892.

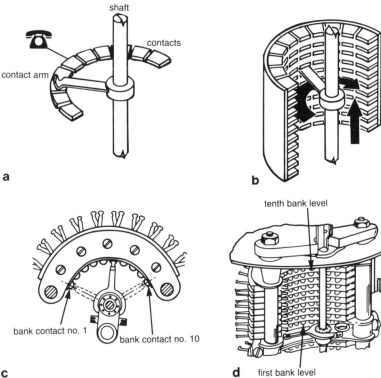

Figure 4.6. Almon Brown Strowger and his automatic telephone switching system. (a) Single-level switch; (b) 10-level switch; (c,d) modern Strowger switch.

The first Strowger exchange in the United Kingdom was opened at Epsom, Surrey, in 1912. By 1918 it had become the norm for automatic exchanges in the United Kingdom.

Strowger lacked the engineering knowledge fully to develop and exploit his invention but with associates the Automatic Electric Company was formed. He died in 1902 but the Company continued with the further development and large-scale manufacture of automatic switching equipment based on the Strowger switch until 1936.

The Bell Telephone System was at first slow to adopt Strowger's switch—perhaps the "not invented here" syndrome may have been in part responsible—but when convinced of the economic advantages of automatic operation, and confident that satisfactory quality of service was achievable, Bell began to buy in quantity from the Automatic Electric Company. In 1916 Western Electric, Bell's manufacturing arm, bought manufacturing rights for Strowger switches and commenced its own production in 1926.[4]

From its base in the United States the Strowger automatic switching system spread all over the world, and although several other systems of automatic switching based on electromechanical techniques were developed in the United States and Europe, up to the 1970s the Strowger system was still the most widely used (see Chapter 14 and Ref. 5).

REFERENCES

1. Robert V. Bruce, *Alexander Graham Bell and the Conquest of Solitude* (Victor Gollancz, London, 1973).
2. William Aitken, *Who Invented the Telephone?* (Blackie, Glasgow/London, 1939).
3. Sydney H. Aronson, "The Sociology of the Telephone," *International Journal of Comparative Sociology, 12,* No. 3, September 1971.
4. *A History of Engineering and Science in the Bell System: The Early Years (1875–1925)* (Bell Telephone Laboratories, 1975), Chapter 6.
5. R. J. Chapuis, *100 Years of Telephone Switching: 1878–1978* (North-Holland, Amsterdam, 1982).

5

Inventors of the
Thermionic Valve

THE CHALLENGE OF LONG-DISTANCE WIRE
AND RADIO COMMUNICATION

As the telephone line networks in the United States and Europe expanded to include intercity transmission over ever-longer distances, problems of unsatisfactory reception related to attenuation of the line current and poor signal-to-noise ratio began to be apparent. At first these difficulties were fought by using heavier-weight, and therefore lower-attenuation, copper conductors, but this approach became impracticable and uneconomic beyond a certain distance. There was also the problem of poor transmission quality resulting from the greater attenuation of the higher-frequency components of the transmitted voice signals compared with the low frequencies. The ideas initiated by Heaviside on inductive loading of telephone cables, developed by Michael Pupin and by George Campbell of the American Telephone and Telegraph Company, improved transmission quality but left unsolved the problem of overall line loss. Early attempts at a repeater/amplifier involved mechanically coupling an electromagnetic receiver and a carbon granule microphone in a battery circuit—a device that produced amplification but yielded poor quality. It was clear that what was needed was some means of amplifying the voice signals without distorting them.

And as radio evolved beyond the sending of Morse code telegraph signals by spark transmitters to voice transmission, so the need arose for means for generating continuous radio-frequency carrier waves and modulating voice signals onto them—techniques that were also essential for multichannel frequency-division wire and cable carrier systems.

The pioneering work needed to create a device that could amplify, generate, and modulate was carried out largely by three men: an Englishman, an Austrian, and an American. The device they evolved was the "thermionic" valve, so called from the Greek *thermos*, meaning warm, which referred to the generation of electrons by a heated wire. American terminology referred to it, perhaps less descriptively, as a "vacuum tube."

The thermionic valve, which was eventually used on a worldwide scale in telecommunications and broadcasting, held the field from the early 1900s until the arrival of the transistor in the 1950s.

John Ambrose Fleming (1849–1945)

John Ambrose Fleming was born in Lancaster in 1849. He graduated from University College, London, in 1870 and entered Cambridge University in 1877 where he worked on electrical experiments under James Clerk Maxwell in the Cavendish Laboratory. He later became Professor of Electrical Engineering at University College, where he carried out experiments in wireless telegraphy and cooperated with Marconi in the design of a transmitter for the first radio transmissions across the Atlantic in 1901.[1,2]

Fleming discovered in 1904 that the electron flow in a vacuum tube with a heated wire cathode was unidirectional toward an anode held at a positive potential relative to the cathode. He realized that such a device could be used to generate a direct current when an alternating potential was included in the circuit, i.e., it would act as a "rectifier." Furthermore, this action was effective even for the high-frequency signals involved in radio waves, and could be used as a "detector" enabling a telegraph or voice signal to be extracted from a modulated radio carrier wave. As such it was a more stable and efficient device than the crystal detectors then in use. Although Edison as long ago as 1883 had

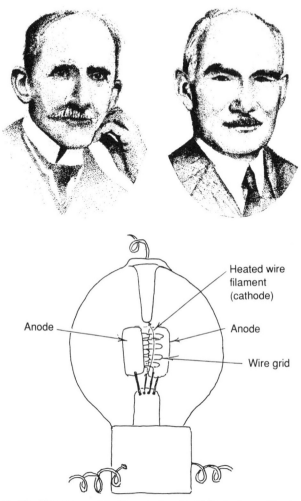

Figure 5.1. (Top) Inventors of the thermionic valve: John Ambrose Fleming (left) and Lee De Forest. (Bottom) Lee De Forest's "audion."

observed an effect involving electrons, it fell to Fleming to realize its potential and to apply it effectively.

Fleming's device became a "diode" thermionic valve, since it had two electrodes. The Fleming diode could not act as a generator or amplifier of electrical oscillations; its function was essentially that of a detector in a radio receiver or a rectifier for the generation of direct current from an alternating current power supply.

But the Fleming diode served another valuable purpose: it triggered the next and most important development of the thermionic valve that enabled it to amplify and generate oscillations.

In 1929 Fleming was knighted for his pioneering work in wireless telegraphy, of which the diode was a part; he lived to the age of 96 and was able to see much of his work come to fruition and wide application.

Robert von Lieben (1878–1914)

Robert von Lieben, an Austrian physicist working in Vienna on the amplification of telephone signals, made the next important step forward. In 1906 he added a third electrode to Fleming's diode, in the form of a wire mesh "grid" between the electron-emitting cathode and the positive potential anode, his device then becoming a "triode."

By varying the negative potential of the grid, the current from cathode to anode could be varied, and if an alternating signal voltage was impressed on the grid, the anode current varied correspondingly and an amplified signal voltage appeared across a resistance in the anode circuit.

This invention was clearly of major importance, opening the door as it did not only to amplification, but also to the generation of stable high-frequency oscillations by coupling together circuits connected to the anode and the grid.

However, von Lieben's work was not fully exploited, perhaps because of his early death at the outbreak of war in 1914, and it was an American, Lee De Forest, to whom most of the credit for development of the triode thermionic valve should go.

Lee De Forest (1873–1961)

Lee De Forest was born in 1873 in Council Bluffs, Iowa. He studied mechanical engineering at Yale University and there became interested

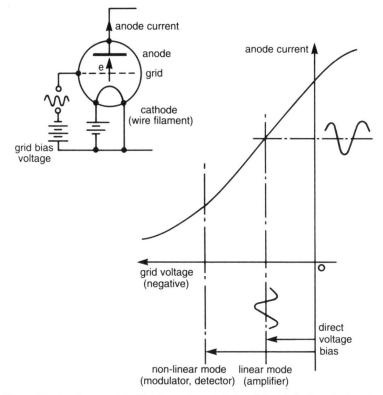

Figure 5.2. Anode current/grid voltage characteristics of the triode thermionic valve.

in radio waves and the new developments in wireless telegraphy, receiving a Ph.D. in 1899 for his thesis on this subject. In 1901 he devised, in competition with Marconi, means for speeding up the sending of wireless signals and was able to interest the U.S. Army and Navy in his work, which found application to news reporting in the Russo-Japanese War of 1904–1905.[3]

He was a prolific inventor, being granted more than 300 patents, but his U.S. Patent No. 841,387 on a "Device for Amplifying Feeble Electric Currents," issued in January, 1907, was the most important. It closely paralleled von Lieben's 1906 discovery in its use of a control grid; however, in De Forest's 1907 patent this was to one side of, rather than across, the electron stream from the cathode, a limitation removed in his 1908 patent on "Space Telegraphy," which showed the grid as a wire mesh through which the electrons flowed on their way to the anode, thus giving more effective control. This device he called an "audion."

At first the De Forest audion was regarded as a very sensitive detector of radio waves, and—perhaps because of the inventor's preoccupation with wireless telegraphy—its potential applications as an amplifier and oscillator were for some five years overlooked, "even by the inventor"![4] It was John Stone Stone, who worked in the Boston laboratory of American Bell, who realized the potential of the audion as an amplifier in the growing telephone network.

Applications as modulators and generators of carrier waves followed, notably by engineers of the Bell System, including H. D. Arnold, R. V. L. Hartley, E. H. Colpitts, and R. A. Heising; these are described in H. J. van der Bijl's book *The Thermionic Vacuum Tube and Its Applications.*

Eventually De Forest sold his "glass bottle full of nothing!" to the American Telephone and Telegraph Company for $390,000—a vast sum in those days but a clear indication of the value of this pioneering achievement.

De Forest was a man of many interests. He staged the first radio broadcast in history from the Metropolitan Opera House, New York, in 1910, and in 1916 set up his own radio station, broadcasting news as well as other programs. In 1923 he became interested in sound motion pictures and demonstrated one of the first "talkies," doubtless using his audion as amplifier!

His life, before the success of his audion, was eventful and hazardous: he, like many inventors, was not a very successful businessman and he lost fortunes as well as making them. He was even placed under arrest at one time for allegedly attempting to use the mails to defraud—this in an effort to finance one of his many inventions.

From these early beginnings many improvements in the design of thermionic valves followed, to which the fundamental studies in thermionic emission by O. W. Richardson (c. 1901) provided a scientific base. Improved vacuum techniques enabled residual gases to be removed and stabilized performance, while the introduction of indirectly heated cathodes by Whenelt (c. 1904) gave greater freedom of circuit design. E. H. Armstrong (see Chapter 8) was one of the first to use the thermionic valve as a circuit element providing amplification, with positive feedback to increase the gain and, where needed, oscillation.

The invention of the thermionic valve was the essential key that opened the door to a wide range of developments in telecommunications and broadcasting. Not only did it overcome the problem of distance, it made possible multichannel carrier cable and radio systems that provided economically for the vast growth of intercity telegraph and telephone traffic that began in the early 1900s. Without the thermionic valve, sound and television broadcasting may well have been delayed until the advent of the transistor in the 1950s. And even today the thermionic valve has a continuing role in the high-power stages of broadcasting transmitters.

REFERENCES

1. J. A. Fleming, *The Principles of Electric Wave Telegraphy* (Longmans, Green, New York, 1906).
2. Ambrose Fleming, *Memoirs of a Scientific Life* (Marshall, Morgan and Scott, London and Edinburgh, 1934).
3. *Pioneers in Telecommunications* (British Telecom Education Service, BT HQ, 81 Newgate St., London, 1985).
4. *A History of Engineering and Science in the Bell System: The Early Years (1875–1925)* (Bell Telephone Laboratories, 1975), p. 256.

6

The Telegraph–Telephone Frequency-Division Multiplex Transmission Engineers

THE DEMANDS OF TRAFFIC GROWTH AND THE SOLUTION

Just as the growth of telegraph traffic and the pressure of economic factors led the early telegraph engineers to develop duplex and quadruplex telegraph systems capable of transmitting two or four messages simultaneously on a single wire, so the growth of intercity telephone communication created demands for providing ever more telephone channels on a pair of wires or a single coaxial cable. This was paralleled by the need similarly to provide more than just two or four telegraph channels on a single transmission path.

The solution to this problem came with the evolution of the concept of "frequency-division multiplexing" in which each telegraph or telephone signal is modulated on a carrier wave of different frequency, transmitted over a common transmission path, selected at the receiving end by a "wave filter" responsive only to a limited band of frequencies, and the telegraph or telephone signal recovered after demodulation of the carrier wave.

An early approach to this concept was Alexander Graham Bell's "harmonic telegraph" (U.S. Patents 161,739 of 1875 and 174,465 of 1876) in which vibrating reeds at the sending end, each tuned to a different frequency, pulsed an electric line current at the same frequency. This line current energized a similarly tuned reed at the receiving end. By such means Bell hoped to transmit several telegraph messages simultaneously over the same line, using what was a primitive form of frequency-division multiplexing based on mechanical, rather than electrical, resonance.

Similar ideas using mechanical resonance were suggested in the 1870s and 1880s by Gray, van Rysselbergh, Edison, and others.

However, the most important advances came when electrical resonance, using tuned circuits with inductance and capacitance, was proposed in the 1890s for multichannel telephony by Michael Pupin at Columbia University, John Stone Stone in the United States, and Hutin and Leblanc in France.

James Clerk Maxwell in England had noted, as early as 1868, the analogy between electrical resonance in a circuit containing inductance and capacitance and the then already well-known mechanical resonance in a system involving a mechanical mass and a spring.

The basic elements of a carrier telephony system involved:

1. A source of stable continuous high-frequency electromagnetic oscillations, i.e., a carrier wave
2. Means for varying the amplitude of the carrier wave in sympathy with the amplitude of a voice wave or telegraph signal, i.e., modulation
3. A wave filter capable of selecting the modulated carrier wave from others of different frequency
4. Means for extracting the desired voice or telegraph signal from the modulated carrier wave, i.e., demodulation

The triode valve—derived from De Forest's "audion"—not only fulfilled the function of a line amplifier by operating on the linear portion of its grid voltage/anode current characteristic, the nonlinear portion enabled it to be used as a modulator or a demodulator, as was demonstrated by van der Bijl and Heising of the Western Electric Engineering Department in the United States.[1]

The later addition of a second grid between the control grid and the anode further improved the performance of the valve by removing unwanted capacitance between the control grid and anode that would otherwise cause unwanted oscillations, thereby inhibiting its use as a high-frequency carrier amplifier.

The development of the thermionic valve received an early impetus from wireless telegraphy and telephony, where it replaced cumbersome and inefficient spark generators and high-frequency alternators, and provided a more sensitive and stable alternative to crystal detectors. In fact, the first multichannel carrier telephony systems on wire were sometimes referred to as "wired wireless" systems.

Although modulation and demodulation in wire and cable carrier systems were later carried out by copper-oxide rectifiers, the thermionic valve remained in use in such systems as audio-frequency and carrier amplifiers until the advent of the transistor in the 1950s.

By 1912 John Stone Stone, an independent worker formerly with American Bell, had written,

> A new art has been born to us. The new art of high-frequency multiplex telegraphy and telephony is the latest addition to our brood of young electric arts.[2]

But much remained to be done before the system design and technology were adequate for commercial use.

The next two important and innovative steps in the development of frequency-division multiplex carrier systems arose not from direct observation of any physical effects but from the theoretical and mathematical studies of engineers in the laboratories of Western Electric and the American Telephone and Telegraph Company in about 1914. One was the creation of a modulation theory that demonstrated quantitatively the characteristics of the sidebands of an amplitude-modulated carrier wave, and the second the principles on which could be designed wave filters sufficiently selective to separate the sidebands.

THE SIDEBANDS ARE REVEALED

J. R. Carson

It seems remarkable now that in the 1920s there were some, including the eminent scientist Sir Ambrose Fleming, who doubted the objective existence of the sidebands of a modulated carrier wave, regarding them as a convenient mathematical fiction and preferring instead to use the "decrement" (or damping factor) of a tuned circuit for defining its response to a modulated carrier wave.

The first clue came in 1914 from the laboratory notebook of a young Western Electric engineer, C. R. Englund, which showed the geometrical relationship of the sidebands and the carrier of an amplitude-modulated wave.[3]

But it was the mathematical analysis of J. R. Carson in 1915 that gave precision to this concept and revealed possibilities for major improvements in the carrier multiplex transmission of telephony. He showed that if a carrier of frequency f and a voice wave of frequency v were passed through a nonlinear device, the output contained not only the original frequencies f and v but also an upper sideband $f + v$, and a lower sideband $f - v$. He further demonstrated that if the carrier and its sidebands were passed into another nonlinear device, the output of the latter contained the original voice frequency v, together with higher-frequency unwanted components that could be removed by filtering. The nonlinear device could, for example, be a thermionic valve operated on the curved part of its grid voltage/anode current characteristic.

Another, important deduction from Carson's analysis was that it was sufficient to transmit only one of the two sidebands to achieve satisfactory voice quality, the carrier suppressed before transmission being replaced by another of the same frequency (or nearly the same within a few cycles per second). This became known as the "suppressed-carrier, single-sideband" mode of transmission. Since the sideband embodied a frequency shift from the original voice frequency, a number of sidebands, each corresponding to a different voice channel, could be stacked side by side in the wide frequency band available on the wire or cable transmission path.

In order to suppress the carrier, a balanced modulator could be used but the selection of a desired sideband required more than a simple tuned circuit; a "wave filter" providing a uniform response over the frequencies of the sideband and rejection of other frequencies was needed. Wave filters were also required to extract a desired sideband from the assembly of sidebands—the multiplex—at the receiving end of the line. Here too, engineers and mathematicians of Western Electric and the American Telephone and Telegraph Company provided solutions.[1]

THE SIDEBANDS ARE SELECTED

G. A. Campbell and O. J. Zobel

G. A. Campbell had worked on the inductive loading of transmission lines to improve their frequency response and in 1910 turned his attention to the design of electric wave filters. Using inductors and capacitors in various circuit configurations, he devised low-pass, high-pass, and band pass filters that could accept frequencies in a pass range and reject to any desired degree frequencies outside that range. The sharpness of transition between the pass range and the reject range, and the degree of rejection outside the pass range could be controlled by suitable circuit design, the circuit requiring more inductor and capacitor elements as the desired performance became more stringent. For voice transmission the pass range of frequencies ranged between about 200 and 2500 Hz in early carrier systems, and eventually increased to 100 to 3400 Hz with 4000-Hz spacing between the virtual carriers.

O. J. Zobel joined the Engineering Department of AT&T with a doctorate from the University of Wisconsin; his mathematical skill, combined with outstanding engineering insight, brought the highly esoteric art of wave filter design to new standards of performance. By 1920 he had evolved filter design techniques that improved the impedance/frequency characteristic within the passband, thereby minimizing reflections, and enabled high peaks of attenuation to be provided outside but close to the passband, thus sharpening the transition region.[4]

These filter developments, to which R. S. Hoyt of AT&T also contributed, had substantial economic value by enabling more sideband voice channels to be packed into the frequency band available on a given wire or cable system.

BENEFITS OF THE SINGLE-SIDEBAND, SUPPRESSED-CARRIER MODE

The effects of Carson's work on modulation theory were both profound and widespread.

One of the first applications of single-sideband operation was to the long-wave 60-kHz transatlantic radio telephone in 1926, where the frequency space available in that part of the radio spectrum was limited.

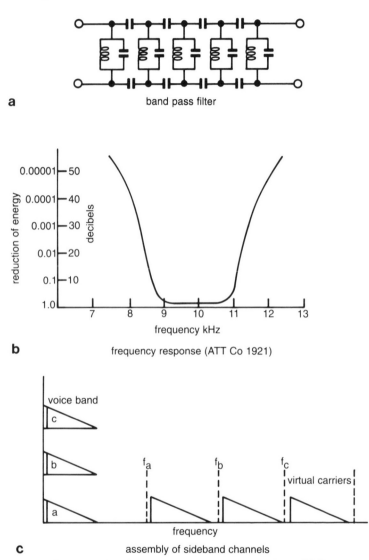

Figure 6.1. Frequency-division multiplexing. (a) Bandpass filter; (b) frequency response (AT&T, 1921); (c) assembly of sideband channels.

The absence of a transmitted carrier enabled the whole of the transmitter output power to be concentrated in the sideband, thereby achieving a substantial improvement in signal-to-noise ratio at the receiver.

The single-sideband, suppressed-carrier mode conferred substantial advantages in multiplex carrier systems. It doubled the number of channels that could be accommodated in a given frequency band on the wire or cable, and greatly reduced the signal power loading on repeater/amplifiers by removing the carrier power, thus minimizing intermodulation crosstalk between channels (Chapter 10).

These advantages were also available on shortwave radio telephone systems, with the added benefit of reduced distortion resulting from multipath radio transmission; many long-distance point-to-point commercial radio links were converted from double-sideband to single-sideband operation from the 1930s onwards (Chapter 7).

Single-sideband, suppressed-carrier techniques became the dominant feature of virtually all wire, cable, and microwave radio-relay frequency-multiplex transmission systems until the 1980s, when digital techniques began to be used (Chapter 13), a remarkable tribute to the far-reaching influence of Carson's original mathematical analysis.

REFERENCES

1. *A History of Engineering and Science in the Bell System: The Early Years (1875–1925)* (Bell Telephone Laboratories, 1975), p. 277 et seq.
2. John Stone Stone, "The Practical Aspects of the Propagation of High-Frequency Electric Waves along Wires," *Journal of the Franklin Institute,* October 1912.
3. J. R. Carson, *Electric Circuit Theory and Operational Calculus* (McGraw–Hill, New York, 1926).
4. T. E. Shea, *Transmission Networks and Wave Filters* (Bell Telephone Laboratories, 1929).

7

Pioneers of Radio Communication

THE TELEGRAPH/TELEPHONE WITHOUT WIRES

By the 1890s the theoretical predictions of Clerk Maxwell and the experimental demonstration by Heinrich Hertz of the existence of electromagnetic "radio" waves had begun to stimulate speculation that perhaps these "Hertzian" waves might be used instead of wires to transmit telegraph and telephone signals over large distances. After all, light—which also consisted of electromagnetic waves—had already been used in the semaphore system to transmit telegraph signals (Chapter 2).

There is no doubt that the first major impetus toward the realization of this dream was given by Guglielmo Marconi's experiments and demonstrations from 1894 onwards, but to which other workers in the United Kingdom, France, and the USSR had made significant contributions.

Up to about 1910 the main emphasis was on the use of radio waves to transmit telegraph signals, but with the invention of De Forest's "audion" triode thermionic valve in 1907, which made possible the generation and modulation of carrier waves (Chapter 5), radiotelephony began to be studied. Despite the differences in the transmission media, the technologies of wire and radio began to merge, including the use of carrier multiplex and single-sideband principles (Chapter 6).

The successful evolution of radio communication in the century beginning in 1890 has involved the detailed study of the propagation of radio waves over an ever-widening frequency spectrum ranging from about 30 kHz (10,000-meter wavelength) to at least 300 GHz (1-millimeter wavelength), and the creation of matching technologies to enable exploitation by a wide variety of radio-based communication services. These include point-to-point and mobile communication involving land, ship, and air stations, sound and television broadcasting (Chapters 8 and 9), microwave intercity links (Chapter 11), satellite communications (Chapter 15), and "cellular" mobile radio communication (Chapter 20).

Let us return to the pioneers who made it all possible.

THE FIRST RADIOTELEGRAPH EXPERIMENTS

Guglielmo Marconi (1874–1937)*

Guglielmo Marconi, the second son of a well-to-do Italian landowner, was born in 1874 in Bologna, Italy. For the most part the boy was brought up at the family's villa near Bologna, but received some of his early education from the age of five at a private school in Bedford, England. On his return to Italy he studied first at a school in Florence, and later completed his formal education at Livorno Technical High School where he studied physics.

From an early age, Marconi displayed an original and inventive mind. In his teens he came into contact with the renowned Professor Righi and studied his work on electromagnetic radiation. A critical moment in his life came when, on holiday in the Italian Alps, he chanced on a paper describing Hertz's scientific experiments, and from this point was fired with the intention of finding a way of using Hertzian waves for communication. He was by no means a scientist looking for explanations of electromagnetic wave phenomena, but rather a practical exploiter of radio waves for communication at a distance.

---

*This account of Marconi's work is based on Ref. 1.

Figure 7.1. Guglielmo Marconi and his first transmitter (1895).

His early experiments were conducted at the family villa near Bologna, where his apparatus was similar to that used by other experimenters of his day, i.e., an induction coil and battery transmitter capable of generating high-voltage sparks across a short air gap in a Hertzian dipole radiator, and a receiver comprising a similar dipole coupled to a "coherer." The latter had been invented earlier by Professor E. Branly; it comprised a vacuum tube with two metal plugs and metal filings between the plugs. It had been christened "coherer" by Sir Oliver Lodge because the metal filings tended to cohere in the presence of an electromagnetic wave train, thereby lowering the electrical resistance between the plugs and enabling current to flow in a battery circuit. The latter thus indicated the presence of the wave train, i.e., functioned as a "detector." To regain its sensitivity after the wave train had ceased, the coherer required gentle tapping, provided by a trembler bell mechanism.

The ranges achieved by Marconi at this time were no more than a hundred yards or so and he left these experiments to study the electrical discharges generated by thunderstorms, using an elevated aerial—as the Russian physicist Popoff had been doing at the time.

On returning to his Hertzian wave experiments, Marconi had the inspiration to use elevated aerials at the transmitter and receiver, each comprising a raised metal plate and a second plate on the ground, in place of the Hertzian dipoles.

The improvement in range was magical—up to a mile and a half. Encouraged by this success, Marconi then set about making detailed improvements to the receiver, added a Morse key at the transmitter and a relay-operated Morse printer at the receiver, and thus enabled telegraphic messages to be transmitted. He was now in business as a communicator.

EARLY ATTEMPTS AT COMMERCIAL EXPLOITATION

Marconi's first effort at commercial exploitation was to offer a demonstration to the Italian Government; it was a bitter blow when, after the usual delay, the Government declared itself not interested. After a family conference it was decided that Marconi, then a young man of 21, should go to England where the family had connections.

There was another, and more powerful, reason for turning to England. He had early realized that one of the most valuable uses of his invention was by shipping—ships in distress had at that time no means of calling for assistance, either from land or another ship. And England in the 1900s had the largest merchant fleet and most powerful navy in the world.

His entry into England was hardly auspicious: an overzealous customs officer had broken his equipment. However, it was soon repaired and on June 2, 1896, Marconi applied for the world's first patent for wireless telegraphy (British Patent 12,039), granted on March 2, 1897. It is questionable whether the term "invention" is strictly applicable to Marconi's system, for he had in effect made use of ideas and devices originated by others but had combined and used them to achieve a commercially useful result.

With sound business instinct, Marconi contacted Sir William Preece, Chief Engineer of the British Post Office,[2] and the British War Office, offering demonstrations. These took place in 1896 between the Post Office Headquarters building in St. Martins le Grand and the Savings

Bank building on Victoria Street, London, and also on Salisbury Plain where a range of one and three-quarters miles was achieved. An observer from the Admiralty at the Salisbury Plain tests was Captain H. Jackson (later Sir Henry Jackson and First Sea Lord) who had already succeeded in sending wireless signals between two naval vessels.

Sir William Preece lectured on Marconi's experiments to the British Association; his interest and support were, however, by no means disinterested. The Post Office had been granted a monopoly in 1896 for telegraphic and telephonic communication in England and Preece, as Chief Engineer, saw wireless telegraphy in other hands as a possible usurper of that monopoly. He accordingly gave Marconi every opportunity to demonstrate his system, on the principle that to be forewarned is to be forearmed. Further experiments on Salisbury Plain followed in 1897 with kite- or balloon-supported aerials, and ranges of some four and a half miles achieved.

Meanwhile, Professor A. Slaby of the Technical High School in Berlin, who was also working on wireless telegraphy, had taken great interest in Marconi's experiments—an interest that later resulted in Marconi's greatest rival, the German Telefunken System, formed by an alliance of the German General Electric and Siemens and Halske companies.

Realizing the growing commercial pressures, Marconi arranged for the formation in July, 1897, of his own company to develop his system commercially; initially named "The Wireless Telegraph and Signal Company Ltd.," it later became "Marconi's Wireless Telegraph Company Ltd." and finally "The Marconi Company."

Marconi continued his experiments; one established contact between an Italian cruiser and the dockyard at Spezia over 11 miles away and resulted in a decision by the Italian Navy to adopt Marconi's system. Another test, in cooperation with the British Post Office, established contact between Salisbury and Bath over a distance of 34 miles. Seeking to interest shipping organizations, he showed how a coastal station at Alum Bay on the Isle of Wight, equipped with a 120-ft mast aerial, could maintain contact with pleasure steamers plying to Bournemouth and Swanage over a sea path of some 18.5 miles.

The Alum Bay demonstration was witnessed by Lord Kelvin in 1898; it must have given him considerable satisfaction and support for his electromagnetic wave propagation theoretical studies. It is recorded that Kelvin sent telegrams from the Alum Bay station via the mainland to Scotland, which he insisted on paying for, thereby both supporting Marconi's infant company and challenging the Post Office monopoly!

THE DEVELOPMENT OF "TUNING"—A GREAT STEP FORWARD

The spark transmitters then in use radiated a broad band of frequencies with the result that a receiver attempting to pick up weak signals from a distant transmitter might well be jammed by the stronger signals from a nearby transmitter. Such interference could have disastrous consequences, for example, to a ship in distress seeking help from a distant vessel. And as the numbers and power of transmitters grew, so the likelihood of interference increased. The problem was all the more acute for Marconi because rival systems in Germany and France were seeking a solution that would give them a competitive lead.

The solution to these problems was found using the principle of "tuning," the use of resonant circuits to limit the spread of frequencies radiated by a transmitter and the band of frequencies to which a receiver would respond. An important lead was given by Sir Oliver Lodge's demonstration in 1889 of the principle of "syntony" in which a receiving wire circle overlapping a transmitting wire circle energized from a spark induction coil could be brought to resonance by adjusting its effective length to correspond with that of the transmitter.

Marconi's contribution—enshrined in his famous Patent No. 7777 of April, 1900—was to couple the transmitting aerial to the induction coil, and the receiving aerial to the coherer, via a high-frequency transformer, a tapped inductor in series with the aerial, and a capacitor (Leyden jar) the capacitance of which could be varied to bring transmitter and receiver into resonance. An important further development was that more than one transmitter could then be coupled to the same aerial, as could receivers to the receiving aerial.

The tuning concept in radio eventually developed similarly to the frequency-division multiplex pioneered by Campbell, Zobel, and Carson for line and cable systems (Chapter 6).

With the more important technical problems solved, the "Wireless Telegraph and Signal Company" began a major drive to commercial

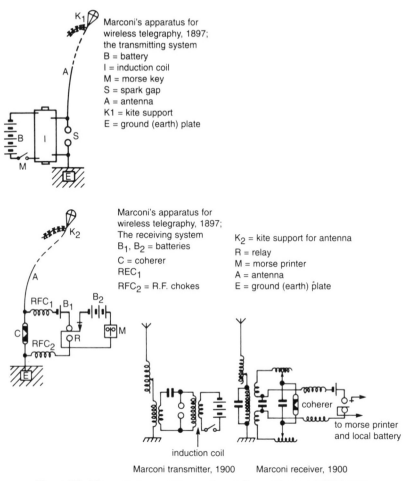

Marconi's apparatus for wireless telegraphy, 1897; the transmitting system
B = battery
I = induction coil
M = morse key
S = spark gap
A = antenna
K1 = kite support
E = ground (earth) plate

Marconi's apparatus for wireless telegraphy, 1897; The receiving system
B_1, B_2 = batteries
C = coherer
REC_1
RFC_2 = R.F. chokes

K_2 = kite support for antenna
R = relay
M = morse printer
A = antenna
E = ground (earth) plate

induction coil

Marconi transmitter, 1900   Marconi receiver, 1900

coherer
to morse printer and local battery

Figure 7.2. Marconi's transmitting and receiving equipment (1897/1900).

exploitation in Europe and the United States, initially with shipboard and coastal radio stations for government authorities such as the British Post Office, British and foreign navies, and private shipping firms.

But a major challenge remained: could the Atlantic Ocean be spanned by wireless telegraphy?

THE ATLANTIC CHALLENGE FOR
WIRELESS TELEGRAPHY

Marconi's attempt to bridge the Atlantic by wireless telegraphy—when the prevailing scientific view deemed this to be impossible—was motivated by a desire to challenge both the submarine cable telegraph companies and the monopoly powers of the British Post Office.

He had little firm evidence to substantiate his belief that so great a distance could be spanned, beyond the fact that some of his earlier experiments had shown that signals could be received some distance beyond the visible horizon. He reasoned that if a transatlantic experiment was to stand any chance of success, it would require the most powerful transmitter that could be built and the highest aerials that were practicable.

Marconi was fortunate in having the services and scientific advice of Dr. J. A. Fleming, then a lecturer at University College, London, who was keenly interested in electromagnetic waves and also had experience in high-voltage alternating currents.

A site was chosen at Poldhu on the Lizard Peninsula, Cornwall, in order to shorten the transatlantic path (the first Post Office satellite earth station at Goonhilly was also sited on the Lizard Peninsula, and for the same reason). The spark-gap transmitter was energized from a 25-kW alternator and keyed by short-circuiting part of the inductance in the alternator circuit. The transmitter power was of course far in advance of the battery-operated transmitters in use up to that time.

The transmitting aerial system at Poldhu, and the receiving aerial at Cape Cod, USA, as first built comprised 200-ft-diameter circles of 20 masts, each 200 ft high, designed by Marconi's assistant Vyvyan.[3] Unfortunately, both aerials were badly damaged by gales in September, 1901. With much fortitude Marconi decided to continue with the work; his assistant Kemp built a temporary aerial system at Poldhu in the form of an inverted triangle of 50 copper wires some 160 ft high. Marconi and Kemp then proceeded to St. John's, Newfoundland, where a receiving system was set up in a hut on the appropriately named Signal Hill, not far from the landing of the first transatlantic telegraph cable. Lacking a suitable mast, the 500-ft-long wire receiving aerial was supported by a kite. The "coherer" in the receiver was one designed for use in the Italian Navy; it comprised a globule of mercury between carbon electrodes, and probably functioned more as a rectifier than the usual iron filing coherer. A telephone-type earpiece was used to listen to the received signals, providing greater sensitivity than was given by the usual relay and Morse printer.

There is uncertainty about the operating wavelength of the system, since wavemeters were not available at that time; one estimate puts it at 366 meters, but it may have been much longer.

The first faint signals from Poldhu, in the form of the three dots of the Morse code for the letter "S," were received on December 12, 1901.

There was at first considerable skepticism about Marconi's claim to have bridged the Atlantic by wireless telegraphy—partly from the mathematical physicists who claimed that, even allowing for diffraction around the Earth, it was impossible and partly because, in the absence of a record from a Morse printer, the claim rested wholly on the unsubstantiated words of Marconi and his assistant Kemp.

And uncertainty was compounded when it was found that the results were not always reproducible, there being a marked difference between the signal strength when the transatlantic path was in darkness and when it was in daylight. It later became clear that an electrified layer in the upper atmosphere, predicted by Heaviside in England and Kennelly in the United States, was playing an important role in guiding the radiated waves across the Atlantic.

Some light was thrown on the wave propagation problem by shipborne receiving tests across the Atlantic on transmissions from Poldhu carried out by Marconi in 1902. These revealed that whereas in daylight the range was limited to some 700 miles, this increased to over 2000 miles at night. This result indicated that there was at least a good prospect of maintaining contact with shipping in the mid-Atlantic both by day and by night.

THE FURTHER DEVELOPMENT OF LONG-WAVE WIRELESS TELEGRAPHY

From 1902 to the 1920s many important developments took place in long-wave wireless telegraphy, and have been described by Vyvyan, Marconi's Chief Engineer.[3]

These included greatly increased transmitter power and the use of continuous waves in place of spark transmissions, at first derived from

Alexanderson high-frequency alternators and later from high-power water-cooled thermionic valves. Receiver sensitivity was gradually improved, at first by using a magnetic detector in place of the coherer, then a point-contact "crystal" rectifier, and lastly the Fleming diode or De Forest triode valve (Chapter 5).

Following objections by the Anglo-American Cable Company to Marconi's operations at Newfoundland, he built, with the agreement of the Canadian Government, a high-power transmitting station at Glace Bay, Canada. He also moved his operations from Poldhu to Clifden on the west coast of Ireland. However, following a disastrous fire in 1909 and troubles in Ireland, this was replaced by a high-power transmitting station at Carnavon, Wales, that commenced public wireless telegraph service in 1914.

European governments, notably in England, Germany, and France, had realized the strategic importance of long-distance wireless telegraphy in time of war and the vulnerability of the undersea cable telegraph system—then the only means of intercontinental communication.

High-power, long-wave wireless telegraphy stations appeared in several countries. In the United Kingdom the Post Office designed and built the long-wave telegraphy transmitter GBR at Rugby; it operated at a frequency of 16 kHz, some 20,000 meters wavelength, with a vast aerial array suspended from 800-ft-high masts and extending over 2 miles. It continues to be one of the the the most powerful radio stations in the world, providing broadcast telegraph press and news services and time signals with a worldwide range extending to the antipodes. This was an interesting application of a principle that Marconi had discovered: the range of long-wave signals guided between the Earth and the Heaviside/Kennelly layer tends to increase as the wavelength is increased.

By 1926 a second high-power transmitter at Rugby, GBY, operating at 50 kHz, 6000 meters wavelength, built by the British Post Office in cooperation with the American Telephone and Telegraph Company, provided a single telephone circuit between the United Kingdom and the United States. It was an early application of the single-sideband technique originated by J. R. Carson and the wave filters due to Campbell and Zobel of Western Electric (Chapter 6).

In the United Kingdom, plans for an "Imperial Wireless Chain"[3] had been made in 1913 with the aim of linking the United Kingdom with Canada, Egypt, India, the Far East, and Australia via a chain of high-power long-wave transmitters, the first of which was to be located

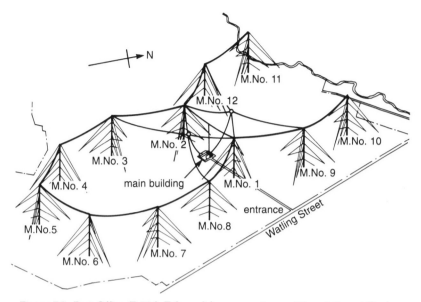

Figure 7.3. Post Office (British Telecom) long-wave transmitting station at Rugby, England. The GBR (16 kHz) aerial system is supported by 12 masts, each 820 ft high. It occupies a site $1^1/4$ miles long and $3/4$ of a mile wide.

at Leafield, Oxfordshire, where the Post Office already had a long-wave transmitter handling its ship–shore telegraph traffic.

However, before the Imperial Wireless Chain could be completed, a new technique of long-distance radio communication became possible—shortwave beamed radio communication via ionospheric reflection—that offered significant advantages compared with long waves. One of Marconi's protégés, C. S. Franklin, played a major role in ushering in the shortwave era.

THE ERA OF BEAMED SHORTWAVE RADIO COMMUNICATION

C. S. Franklin

The limitations of the long-wave radio system were all too apparent: for long-distance transmission the aerial systems required were massive,

Figure 7.4. C. S. Franklin, father of the shortwave beam array.

the transmitter powers needed were large, and the space available in the frequency spectrum precluded more than a limited number of telegraph services, with little room for telephone services. Moreover, it was impracticable to concentrate transmitted wave energy into a beam since the aerial dimensions required would have been excessive.

Hertz's experiments in 1886 had already demonstrated that short-wavelength radio waves could be concentrated into a beam by parabolic mirrors (Chapter 2), and this no doubt led Marconi to arrange for C. S. Franklin to carry out an intensive study of the propagation characteristics of shortwaves and their possibilities for telegraph and telephone communication.

By 1919 thermionic valves capable of generating a few hundred watts of continuous wave power at wavelengths down to about 15 meters had been developed and were used initially for overland tests in England that achieved ranges of up to about 100 miles with the aid of parabolic reflector aerials.

It was then decided to carry out a trial of larger scale, using some 10 kW of power on a wavelength of 100 meters and a parabolic aerial mounted 325 ft high on a mast at Poldhu. The receiver, which did not use a reflector at the aerial, was mounted on Marconi's yacht *Elettra*. By sailing into the South Atlantic it was shown that ranges of 1250 nautical miles by day and at least 2230 miles by night were achievable. Other tests in 1924 using a wavelength of 100 meters with a power of about 20 kW at Poldhu reported good nighttime reception in New York, and in May of that year speech was for the first time successfully transmitted from England to Australia.

It must in fairness be pointed out that radio amateurs, who had been given wavelengths below 200 meters because they were thought to be useless for commercial purposes, also began to demonstrate long-distance contacts, often with powers of only a few watts and aerials of limited directivity.

Instead of the somewhat cumbersome parabolic reflectors used in the early experiments, Franklin devised a highly efficient "curtain array" directional aerial system based on half-wavelength co-phased wire elements arranged in vertical and horizontal groups, and linked to the transmitter or receiver by nonradiating low-loss coaxial feeders. This mechanically convenient, efficient, and economic solution to aerial design contributed substantially to the development of shortwave radio communication.

These results gave Marconi and his team confidence to propose to the British Government an "Imperial Beam System" based on beamed shortwaves to replace the earlier planned long-wave high-power system—offering lower costs, better performance, and greater flexibility. The Government, to avoid a conflict of interest between cable and wireless companies and to ensure more effective utilization of both media, brought both together in a new company Imperial and International Communications Ltd., later named Cable and Wireless Ltd., with the obligation to provide worldwide telegraphic communication. In the United Kingdom shortwave transmitting stations equipped with Franklin-type aerial arrays were built at Ongar and Dorchester, and receiving stations at Brentwood and Somerton. These stations worked to similarly equipped overseas stations, many supplied by, or built under license from, the Marconi Company.

Development of Beamed Shortwave Radio Communication
by Engineers of AT&T

Recognizing the limitations of a single-channel transatlantic radio-telephone link using long waves (60 kHz), and noting the encouraging results from Marconi's shortwave tests from Poldhu, England, engineers of AT&T, Western Electric, and Bell Laboratories in the United States began to explore the possibilities of shortwave radio in the 1920s. An experimental transmitting station was established at Deal, New Jersey, and the development of high-power transmitters put in hand, so successfully that by 1930 a 60-kW output stage using water-cooled valves and driven by linear lower-power radio-frequency amplifiers had been achieved. Shortwave receiving stations were set up at Holmdel and Netcong, New Jersey.[4]

Prominent among the engineers who made important contributions to this work were:

A. A. Oswald
J. C. Schelleng
M. J. Kelly
H. T. Friis
W. Wilson
E. J. Sterba
R. A. Heising
E. Bruce

Considerable effort involved the development of shortwave arrays in view of the advantages of the gain in effective radiated power and the reduction of interference offered by the array directional properties. One design—the Sterba array—was similar to the Franklin vertical curtain of half- and quarter-wave elements. Another—the Bruce array—at Netcong used a horizontal group of vertical half-wave elements with a similar reflecting group a quarter-wavelength away.[5]

A steerable version of the Bruce array was used by K. G. Jansky in his discovery of interstellar radio-frequency noise radiation in the Milky Way.[6]

E. Bruce, with H. T. Friis, was responsible for the most widely used of all shortwave arrays, the horizontal rhombic aerial, which largely displaced the large curtain arrays for commercial point-to-point radio services.

The horizontal rhombic aerial had the substantial advantages of simplicity and low cost: it could be supported on poles some 80 ft high, in contrast to the expensive structures needed to support the vertical curtain arrays. Moreover, it was effective over a frequency range of about 4 to 1, e.g., from 5 to 20 MHz, as compared with the relatively narrow bands offered by resonant element systems; effective power gains of some 14 decibels could be realized.

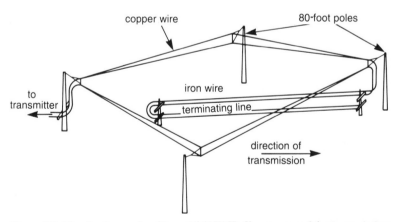

Figure 7.5. The shortwave rhombic aerial (AT&T). Shown as used for transmission; also used for reception with resistor termination.

From the mid-1920s to the 1930s exhaustive radio-wave propagation tests were made, mainly on the transatlantic path to England where
a receiving station had been established by AT&T at Southgate, near
London, and in cooperation with the British Post Office at their receiving
station near Baldock, Hertfordshire. These tests sought to provide data
to determine the optimum frequencies for use at various times of the
day and seasons of the year, taking into account the varying influence of
solar sunspot activity on the electrified E and F layers of the ionosphere.
Experiments in which multitones spaced at about 100 Hz were transmitted in the voice band clearly revealed the existence of selective fading
resulting from multipath transmission via the various layers and multiple-hop paths.[8]

Developments in Europe, notably in the British Post Office, had
closely paralleled those in the United States. AT&T built a transmitting
station for commercial operation at Lawrenceville, New Jersey, while
the British Post Office established one at Rugby, England. In September
of 1929 a demonstration was given, in conjunction with the British Post
Office, in which Bell Laboratories officials in New York talked with
Australia via London—a historic occasion that marked the beginning
of the shortwave point-to-point radio communication era.

The first commercial radiotelephony services were operated on
a double-sideband amplitude-modulation basis; however, it had been
realized for some time that single-sideband operation offered a potential
9-decibel improvement of signal-to-noise ratio (6 dB from suppression
of the carrier wave, and 3 dB from the reduction of bandwidth by
removal of one sideband), equivalent to a transmitted power increase
of 8 times. An additional advantage was the removal of nonlinear distortion that occurs in double-sideband operation when the carrier fades
relative to the sidebands. The problems of frequency stabilization at the
receiver to enable the reinserted carrier frequency to be corrected within
a few cycles per second were eventually solved and single-sideband
working began to be used generally from the mid-1930s. Further developments enabled four telephone channels to be provided, two on either
side of a partially suppressed carrier, described as "independent-sideband" operation.

The last major technological development in the shortwave era was
the Multiple-Unit Steerable Antenna for Short-Wave Reception (MUSA),
which had the aim of further improving the quality and reliability of
shortwave point-to-point radio communication by separating the rays

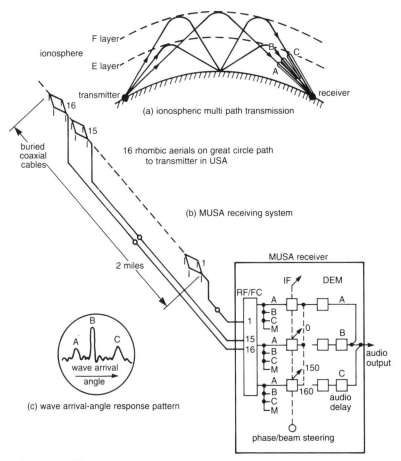

(a) ionospheric multi path transmission

16 rhombic aerials on great circle path
to transmitter in USA

(b) MUSA receiving system

(c) wave arrival-angle response pattern

Figure 7.6. The multiple-unit shortwave steerable array (MUSA) (1940–1960).

due to multipath propagation and then combining the demodulated
voice signals after delay correction. This brilliant concept was the product of the pioneering group of Bell Laboratories engineers at Holmdel,
and has been described by H. T. Friis and C. B. Feldman in Ref. 9. An
account of the matching MUSA designed and built by the British Post
Office for the transatlantic link is given below.

The story of Bell System contributions to the shortwave era given in Ref. 4 ends on a somewhat wistful note:

> A sentimental reunion ceremony was held for the many engineers who contributed to the development of the short-wave art on the closing down of the Lawrenceville (transmitting) station in February 1976. Most of the people had already retired, but history would have been kinder to them if it had taken a bit longer for the short-wave radio-telephony art to become obsolete.

Nevertheless, even today the shortwave radio spectrum is crowded with commercial, broadcasting, and amateur transmissions—a lasting tribute to the pioneers whose work made it possible.

Development of Beamed Shortwave Radio Communication by Engineers of the British Post Office

The Government had placed on the British Post Office the obligation to develop long-distance overseas radiotelephony services to and from the United Kingdom, with Cable and Wireless Ltd. handling the overseas telegraphy services. The equipment development for radiotelephony became the responsibility of the Engineering Department of the British Post Office, which it carried out under the direction of its Engineers-in-Chief, notably G. Lee, C. Angwin, A. J. Gill, and A. H. Mumford, all of whom were knighted for their services. In addition to the long-wave transmitters GBR and GBY referred to above, they were responsible for building shortwave transmitting systems at Rugby and receiving systems at Baldock.

The pioneering work of the British Post Office engineers in the development of beamed shortwave radiotelephony systems from about 1920 to 1950 included the design and construction of:

- Aerial arrays
- High-precision quartz-crystal drives for transmitter and receiver frequency control
- High-power transmitters
- Superheterodyne receivers
- Single-sideband and independent-sideband (four-channel) transmitter drives and receivers

The general development of radio communication in the British Post Office has been surveyed in Ref. 10. In much of this work there was close liaison with engineers of AT&T and the Bell Telephone Laboratories. Some developments such as single- and independent-sideband working were first tested on the transatlantic route and later extended to other overseas routes.[11]

As noted above, the British Post Office in cooperation with AT&T also built a MUSA receiving system for the transatlantic link, described by A. J. Gill in his address as Chairman of the Wireless Section of the Institution of Electrical Engineers, London, in 1939.[12]

The MUSA represented the ultimate development in shortwave receiving systems: it comprised 16 horizontal rhombic aerials arranged on the great circle path pointing toward the distant transmitter, the radio-frequency outputs of which were passed over low-loss coaxial feeders to the receiving installation. There, after conversion to an intermediate frequency, the signals were adjusted to combine in phase. In this way, extremely sharp spatial directivity was achieved in the vertical plane, enabling individual down-coming rays reflected from the E and F layers of the ionosphere to be selected and the demodulated voice signals then combined. The overall effect was to improve the received signal-to-noise power ratio by up to 30 times, together with an improvement of voice quality by minimizing distortion related to radio wave path time-delay differences.

The British Post Office MUSA was located on the Cooling Marshes, Kent; AT&T built their system at Manahawkin, New Jersey. The British system was unique in that it used an electrical phase-shifting system based on an artificial long line with 16 tappings, whereas the American system used a mechanical system with a 16-step gearbox. The British phase-shifting system was in some ways a precursor of the phase-shifting arrays later developed for radar.

Although the expected improvement in transatlantic communication reliability was not always achieved—as for example when the ionosphere was badly disturbed as a result of solar activity—the MUSAs nevertheless gave valuable service from 1940 to 1960. In the absence of submarine cables this service was especially significant during the war years when direct communication between the U.S. and U.K. Governments was vital.

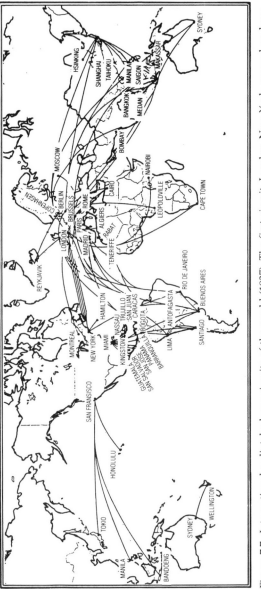

Figure 7.7. International radiotelephone circuits of the world (1937). The first circuit, London–New York, opened on long waves in 1927.

THE ROLE OF THE SHORTWAVE RADIO COMMUNICATION SERVICES

The long-distance shortwave radiotelephone and telegraph services undoubtedly played a vital role in meeting the need for ever-expanding worldwide communications from the 1920s to the 1960s. For this the named, and the many unnamed, pioneers who made it possible deserve our thanks and appreciation. It is easy to forget that sometimes, as in Marconi's early experiments, they were working in a field where scientific knowledge was often lacking and progress had to be made by engineering intuition.

The shortwave services were, however, subject to interruption and loss of voice quality at times of ionospheric storms, and the scope in the frequency spectrum for continued growth of services was itself limited. By the 1950s multichannel telephony submarine cables were coming into service and satellite communications were soon to follow.

The era of long-distance shortwave beamed point-to-point radio communication is coming to a gradual close; nevertheless, for certain applications such as world-service broadcasting, maritime and defense services, shortwave radio has a continuing role.

REFERENCES

1. W. J. Baker, *A History of the Marconi Company* (Methuen, London, 1970).
2. E. C. Baker, *Sir William Preece, FRS, Victorian Engineer Extraordinary* (1976).
3. R. N. Vyvyan, *Wireless Over Thirty Years* (Routledge, London, 1933).
4. *A History of Engineering and Science in the Bell System: Transmission Technology (1925–19750)*, edited by E. F. O'Neill (Bell Laboratories, 1985), Chapter 2.
5. E. Bruce, "Developments in Short-Wave Directive Antennas," *Bell Syst. Tech. J., 10,* October 1931.
6. K. G. Jansky, "Electrical Disturbances Apparently of Extra-Terrestrial Origin," *Proc. IRE,* October 1933.
7. "Horizontal Rhombic Antennas," *Proc. IRE, 23,* January 1935.
8. R. K. Potter, "Transmission Characteristics of a Short-Wave Telephone Circuit," *Proc. IRE, 18,* April 1930.
9. A. A. Oswald, "The Manahawkin MUSA," *Bell Lab. Rec., 18,* January 1940.
10. "The Development of Radio-Communication in the Post Office," *Post Off. Electr. Eng. J., 49,* Pt. 3, October 1956.

11. W. J. Bray *et al.*, "Single-Sideband Operation of Short-Wave Radio Links," *Post Off. Electr. Eng. J.*, *45*, October 1952, January 1953; *46*, April, October 1953.
12. Chairman's Address, A. J. Gill, Wireless Section, Institution of Electrical Engineers, London, *J. IEE, 84*, No. 506, February 1939.

Further Reading

G. Bussey, *Wireless: The Crucial Decade 1924–34*, IEE History of Technology Series No. 13 (London, 1990).
P. Rowlands and J. P. Wilson, "Oliver Lodge and the Invention of Radio," PD Publications, (London, 1994).

8

Pioneers of Sound
Radio Broadcasting

THE BEGINNING OF SOUND RADIO BROADCASTING
IN THE UNITED STATES

Once the problems of generating a continuous carrier wave and modulating it by speech had been solved and applied in point-to-point and
ship–shore services, it was perhaps inevitable that enterprising amateurs and entrepreneurs would seek to explore the possibilities of the
new technology for communication with a wider audience.

The first broadcasting stations appeared in the United States where
the absence of strict government regulation made it possible for anyone
to set up and operate a radio station, requiring little more than the
registration of the wavelength to be used— at first around 360 and 400
meters. Lee De Forest, whose name is associated with the invention of
the triode valve (Chapter 5), was one of the first to set up a low-power
amateur station in 1907.

In 1919 the Radio Corporation of America had come into being
having acquired substantial patent rights from the British Marconi Company, the De Forest patents on the triode valve, and the Westinghouse

patents of Armstrong and Fessenden on the principles of heterodyne reception (use of a nonlinear device to mix a carrier and local oscillation to generate an intermediate frequency) and regeneration (feedback to improve reception sensitivity). RCA was thus in a strong position commercially to exploit sound radio broadcasting, but its success in so doing owed much to an organizing genius, David Sarnoff.

David Sarnoff (b. 1891)

At the age of 15, David Sarnoff entered the American Marconi Company as an office boy in 1906 but soon displayed a keen interest in wireless telegraphy, carrying out experiments in his spare time. He became a ship's wireless operator and performed valuable services at the time of the *Titanic* disaster. By 1916 he had risen to the position of Contracts Manager in RCA and made the then remarkable proposition to the company that transmitting stations be built for the purpose of broadcasting speech and music, and a "radio music box" be manufactured for sale to the general public.

The proposal was revolutionary and had breathtaking implications; the very idea of radio for "entertainment" was new but if the public found it acceptable the mass market for receivers could be enormous and commercially very profitable.

Newspaper owners, manufacturers, and retail store owners were quick to realize the potential of broadcasting for advertising their wares, and by 1920 a boom in broadcasting had started in the United States. One of the most well known was the high-power station KDKA in Pittsburgh, Pennsylvania, built by Westinghouse, partly to stimulate its sales of receiving equipment.

By 1924 there were more than 1000 stations in operation in the United States, and since the few wavelengths then available had often to be shared by several stations there was a great deal of mutual interference. To remedy this chaotic situation, and to impose a degree of control over the utilization of the frequency spectrum, the U.S. Federal Communications Commission was eventually formed.

THE BEGINNING OF SOUND RADIO
BROADCASTING IN BRITAIN

In Britain sound radio broadcasting got off to a slower, but better controlled, start. The British Post Office was responsible to Parliament for the regulation of wireless telegraphy, and this was held to include radio broadcasting. At first the Post Office licensed only low-power transmitters for experimental purposes, but by 1922 had authorized the Marconi Company to carry out limited broadcasting from an experimental station at Writtle, near Chelmsford, with a maximum power of 250 watts.[3]

The Writtle station operated at a wavelength of 400 meters with a four-wire aerial 250 ft long and 110 ft high, the transmitter being adapted from a standard radiotelephone of the time. The station call sign was 2MT, pronounced "Two Emma Toc."

The license for the experimental station permitted only a meager half hour of broadcasting a week, but even this did not deter the small group of Marconi engineers, led by Capt. P. P. Eckersley, who built and operated the station.

Capt. P. P. Eckersley (1892–1963)

P. P. Eckersley and his brother T. L. Eckersley (see p. 105) were grandsons of the great Victorian scientist Professor Thomas Henry Huxley, and there is little doubt that some of the genius of their illustrious forebear "rubbed off" on the Eckersleys.

The first "entertainment" broadcast was made from the wooden hut at Writtle on February 14, 1922, to the joy of the many amateurs who had pressed for a radio broadcasting service. And P. P. Eckersley proved to be not only a competent engineer but also a first-class entertainer, gifted with spontaneous humor that brought enthusiastic responses from the listening audience.

These early, and necessarily brief, programs included not only a rudimentary "Children's Hour"—of which P. P. Eckersley was "Uncle"—but also the first radio play (*Cyrano de Bergerac*), performed via a hand-held carbon microphone passed from performer to performer. And it was from Writtle that Dame Nellie Melba made a historic first operatic broadcast.

Figure 8.1. Captain P. P. Eckersley and the first experimental broadcasting transmitter at Writtle, England (1921).

Figure 8.2. The Writtle pioneers of the birth of broadcasting in the United Kingdom. Left to right standing: B. N. MacLarty, H. L. Kirke, R. T. B. Wynn, and H. J. Russell; seated: F. W. Bubb, P. P. Eckersley, E. H. Trump and Miss F. M. Beeson.

And so, in spite of a lack of official enthusiasm, British sound radio broadcasting was born, mainly through the efforts of a few pioneers with vision and drive. To quote from W. J. Baker:

> On 17 January 1923, the Marconi station 'Two Emma-Tock' at Writtle had made its final bow to a regretful circle of enthusiasts. It closed down with full honours, its mission accomplished . . . it had created an enthusiasm for broadcasting that was destined to make it into the greatest medium for entertainment and instruction that the world had ever seen. In its short and hilarious career it had laid the foundations of the age of broadcasting.

Meanwhile the Post Office had authorized the Marconi Company to set up an experimental station, call sign 2LO, at Marconi House in London with a maximum power of 1.5 kW and a wavelength of 360 meters. Transmission times were at first limited to one hour daily but the programs began to draw sizable audiences by including music concerts and outside broadcasts of special events.

THE BRITISH BROADCASTING COMPANY IS CREATED

As was to be expected, the granting of an experimental radio broadcasting license to the Marconi Company brought many applications from other U.K. manufacturing companies for similar facilities. To avoid the possibility of broadcasting chaos—the signs of which were already apparent in the United States—the Post-Master General invited all interested parties to form a consortium to create a single broadcasting authority for the United Kingdom.

The outcome was the British Broadcasting Company Ltd., formed from Marconi's Wireless Telegraph Co. and five other manufacturers, with revenue derived from a tariff on all broadcast receiving equipment sold by them and half the ten-shilling license fee paid by all listeners.

P. P. Eckersley joined the BBC in 1923, becoming its Chief Engineer until he left the company in 1929. His major achievement was his "Regional" broadcasting scheme in which twin high-power transmitters at each site radiated two programs on different wavelengths.

The growth of new stations was phenomenal: stations in Birmingham, Manchester, Newcastle, Cardiff, Glasgow. Aberdeen, Bournemouth, and Sheffield followed in quick succession.

P. P. Eckersley had a visionary quality that is the stamp of greatness, a charmimg and powerful personality, and a remarkable ability to deflate overpompous individuals. He did more than anyone to put the technical foundations of broadcasting in the United Kingdom on a sound footing. His book *The Power Behind the Microphone*[2] gives a vivid account of the problems—organizational and technical—that he fought to overcome.

However, this brief account of the newly formed broadcasting company would be incomplete without mention of another BBC engineer whose innovative abilities played an important role in its development, Capt. H. J. Round.

Capt. H. J. Round (1881–1966)

H. J. Round studied at the Royal College of Science, London, where he gained a first-class honors degree. He worked initially as an assistant to Marconi on direction-finding and long-wave transmitter design.

Figure 8.3. Radio-communication pioneers of the Marconi Company. (Left) Capt. H. J. Round—circuit and device inventor. (Right) T. L. Eckersley, FRS—outstanding radio-wave propagation theorist.

Before the First World War he became involved in improvements to thermionic valves, including the use of oxide-coated filaments and indirectly heated cathodes, taking out numerous patents in 1913–1914. During the war years he became responsible for setting up direction-finding networks on the Western Front and in the United Kingdom; the latter detected the movement of the German fleet from its base in Wilhelmshaven that signaled to the British Admiralty that the Battle of Jutland was imminent.

As medium-wave broadcasting expanded territorially, the need for higher transmitting power to extend the coverage of individual stations became apparent, and in this area Capt. H. J. Round made important contributions. He also furthered the development of microphones and receivers, including the "reflex" circuit which enabled a single valve to operate simultaneously as a high-frequency and an audio-frequency amplifier.

THE REFLECTING LAYERS OF THE IONOSPHERE ARE REVEALED

As information concerning the performance of medium-wave broadcasting transmitters began to build up, it became apparent that after nightfall the range over which signals could be received was often greatly extended, but in this extended range there was both fading and distortion.

Evidently a "sky wave" was appearing at night, possibly as a result of reflection from an elevated atmospheric layer or layers. This phenomenon also caused interference when transmitters within an extensive geographical area such as Europe shared the same wavelength. Clearly more information as to the nature and properties of the reflecting layers was needed, partly to determine the extent to which wavelengths could be shared but also to enable transmitting aerials to be designed for minimum sky-wave radiation.

Dr. E. Appleton of Kings College, London carried out an important experiment in the 1920s in which the frequency of the BBC transmitter 6BM at Bournemouth was gradually shifted over a small range and the interference fringes between the ground wave and reflected sky wave were observed at a suitably placed receiving station. The interference fringes revealed a reflecting layer at a height of about 100 km, named by Appleton as the "E layer" (not E for "Edgar," Appleton disclaimed, but E for "electron"!).

In the United States, Breit and Tuve of the National Bureau of Standards Laboratory transmitted pulsed signals vertically and demonstrated the multiple-layer structure of the ionosphere. It became clear that there were other layers, named F1 and F2, above the E layer, and an absorbing D layer below it, the behavior of which varied daily, annually, and over an 11-year cycle linked with sunspot activity.

In view of its great value for the planning and operation of radio broadcasting and point-to-point communication systems, the scientific study of radio-wave propagation and the ionosphere became an important responsibility of the Radio Research Station of the National Physical Laboratory in the United Kingdom, and the National Bureau of Standards in the United States.

Important contributions to this study were made in the 1920s and 1930s by T. L. Eckersley of the Marconi Company.

T. L. Eckersley, FRS (1887–1959)

T. L. Eckersley acquired a B.Sc. degree at University College, London, and did research at the Cavendish Laboratory, Cambridge, and the National Physical Laboratory.

During the First World War he worked with the Royal Engineers, carrying out theoretical and experimental radio studies on the "night effect"of sky waves and the effect of coastal refraction on direction-finding. He joined the Marconi Company in 1919 and commenced the intensive study of radio-wave propagation that became his life's work.

With the advent of the Marconi–Franklin beam system of short-wave communication he turned his attention to the propagation of high-frequency electromagnetic waves and during the period 1924–1934 directed a research team that carried out many pioneering investigations of the ionosphere. This work resulted in a number of papers for the Institution of Electrical Engineers that laid down a basis for the prediction of the performance of high-frequency radio services that were brilliantly confirmed in practice.

T. L. Eckersley's genius led him to apply the phase-integral method, familiar in quantum mechanics, both to the magneto-ionic theory of ionospheric propagation and to the diffraction of radio waves around the curved Earth. The work was later extended to include tropospheric-scatter propagation at very high frequencies. In 1938 he was made a Fellow of the Royal Society for these outstanding contributions to the study of radio-wave propagation, and a Fellow of the American Institute of Radio Engineers in 1946. The citation for the latter reads:

> Both his approach to the problem from the standpoint of communications and his invention of mathematical tools useful in the computation of radiated fields are achievements of lasting value, acclaimed by the whole radio world and form a monument of which he may be justly proud.

BROADCASTING ON VERY HIGH FREQUENCIES

Following the end of World War II it became apparent that sound radio broadcasting would need to expand beyond the numbers of transmitting stations that could be accommodated in the long- and medium-wave bands. Attention was then directed toward the very-high-frequency (VHF) band 87.5–100 MHz that had been allocated to broadcasting by international agreement, and which offered much greater scope for growth. It also enabled the quality of reception to be improved by increasing the band of audio frequencies that could be transmitted to 20 Hz–15 kHz needed for high-fidelity music transmission. And interference from distant transmitters could be greatly reduced, partly because of the better radio-wave propagation characteristics of VHF, and partly because directional aerials were possible. However, since the range available did not extend greatly beyond the visible horizon, more transmitters albeit of lower power were needed to provide a service over a given area.

Much of the technology developed during the war years for radar and fighting services radio communication was available or could be adapted for broadcasting.

But an important question remained: what system of modulation should be used for public service VHF broadcasting in the United Kingdom, the well-tried amplitude modulation (AM) used on long and medium waves, or frequency modulation (FM) which by 1945 already had considerable application in the United States?

SOUND BROADCASTING ACQUIRES ENHANCED QUALITY: THE INVENTION OF FREQUENCY MODULATION

Major Edwin Armstrong

In the United States, Armstrong had been led to consider the use of FM of a radio carrier wave in the early days of long-wave radiotelegraphy in an attempt to eliminate static interference. His experiments were not successful and a mathematical analysis by J. R. Carson of Bell suggested that there was no bandwidth reduction or noise improvement to be gained by the use of FM compared with AM.

However, Armstrong pursued his study of FM and, by excellent engineering intuition , introduced the all-important concept of an amplitude limiter in FM reception (this had been omitted in Carson's mathematical model). The effect was almost magical: by arranging that the peak-to-peak frequency deviation of the carrier wave was several times the highest audio frequency to be transmitted, most of the noise background at the FM receiver output fell outside the wanted audio band and a substantial improvement in signal-to-noise ratio was achieved.

An additional bonus was a "capture" effect which caused a weaker interfering carrier to be virtually eliminated.

Armstrong had a long struggle to perfect his new system and to convince American manufacturers and broadcasters of the value of his invention, but by 1945 the widespread use of FM in the United States was proof of its worth.

In Britain, interest in FM was no doubt stimulated by a long-standing friendship between Major Armstrong and Capt. Round of the Marconi Company. After the end of the war the BBC Research Department, headed by H. L. Kirke, the Marconi Company, and the British Post Office began an extensive series of tests to assess the advantages of FM and to optimize the technical characteristics to be used in the public broadcasting system.

By 1950 the first BBC 25-kW FM VHF transmitter, built by the Marconi Company, commenced operation at Wrotham Hill, Kent, and a new high-quality public broadcasting service in the United Kingdom began.[1-3]

FM also had an important role in the VHF mobile radio services used by ambulance, police, and other public authorities and by private users that began to develop in the postwar years.

SOUND BROADCASTING ACQUIRES TWO-DIMENSIONAL REALITY: STEREOPHONIC BROADCASTING AND RECORDING

Alan Blumlein (1903–1942)

Sound quality and realism in radio broadcasting and record reproduction have been greatly enhanced by stereophony, i.e., the reproduced sound appears to originate from sources distributed in space as were the original sources, rather than from a single point in space.

While the research organizations of broadcasting authorities have made useful contributions in this field, there is little doubt that a major and initial impetus came from a British inventive genius, Alan Blumlein.*

---

*A program describing Alan Blumlein's life and work was broadcast by the BBC on Radio 4, January 21, 1990, and has been used in this account.

Alan Blumlein was born in 1903, educated at Highgate School, and graduated from City and Guilds Engineering College, London; his first-class honors degree was in heavy-current electrical engineering. In 1924 he joined Standard Telephones and Cables Ltd. where his early work concerned the improvement of transmission on long-distance telephone lines.

In 1929 he joined the Columbia record company, which later merged with His Master's Voice record company to form EMI (Electric and Musical Industries Ltd.). U.S. record companies had made consider-able improvements in sound recording techniques and Blumlein was asked by the Managing Director, Isaac Schoenberg, to study ways of circumventing the U.S. patents; Blumlein did so and produced even better techniques for driving the record-cutting stylus.

In the 1930s the Bell Telephone Laboratories in the United States had been studying methods for recreating a more realistic sound field using multiple loudspeakers. Recognizing that a simpler approach was commercially desirable, Blumlein concentrated on "deceiving the ear" by a pair of loudspeakers energized from two microphones, each having a directional pattern, and with the pattern axes arranged at 90°—the "crossed figure-of-eight." Blumlein's 1931 patent covers this and other microphone configurations, but even more importantly, means for re-cording pairs of stereo signals in a single record groove or a single film track.

However, it was many years before the record companies and the film industry took stereophony seriously and it is doubtful whether Alan Blumlein ever received an adequate reward for his inventions.

Blumlein's vital contributions to the development of television are described in Chapter 9; in addition to these he carried out important work on airborne radar ground surveillance techniques in the war years. He died in an air crash in 1942 while testing radar equipment. His colleagues described him as at times a difficult man but with a remark-able ability to cross boundaries between scientific philosophies and between groups of people.

A detailed account and appreciation of Blumlein's work, and an illuminating pen-picture of his personal qualities, has been given by Professor R. W. Burns in Ref. 4:

> A. D. Blumlein (1903–1942) was possibly the greatest British electronics engineer so far this century. By the time of his death in 1942 at the age of

38 he had been granted 128 patents—an average of one for every six weeks of his working life—testimony to an inventive genius who made a major impact on the fields of telephony, electrical measurements, sound recording, television and radar.

I. Schoenberg, General Manager of EMI, said of Blumlein shortly after his death:

There was not a single subject to which he turned his mind that he did not enrich extensively.

A further series of contributions on Blumlein's life and work appears in Ref. 5.

REFERENCES

1. W. J. Baker, *A History of the Marconi Company* (Methuen, London, 1970).
2. P. P. Eckersley, *The Power Behind the Microphone* (Jonathan Cape, London, 1941).
3. Asa Briggs, *The History of Broadcasting in Great Britain:* Vol. 1, *The Birth of Broadcasting* (Oxford University Press, London, 1961).
4. R. W. Burns, "A. D. Blumlein—Engineer Extraordinary," Institution of Electrical Engineers, London, Engineering Science and Education Journal, February 1992.
5. R. W. Burns, A. C. Lynch, J. A. Lodge, E. L. C. White, K. R. Thrower, and R. M. Trim, "The Life and Work of A. D. Blumlein," Institution of Electrical Engineers, London, Engineering Science and Education Journal, June 1993.

9

Pioneers of Television Broadcasting

When one views today's superb high-quality color television pictures—with the prospect of even higher definition, large-screen receivers offering three-dimensional pictures to come—it is difficult to realize how crude were the early attempts to achieve "viewing at a distance." It is very doubtful whether even the most optimistic and farsighted of the pioneers had any clear vision of what, in the fullness of time, was to be achieved.

The history of the development of television is a fascinating story of numerous small advances by many individuals toward the ultimate goal, some brilliant ideas by a few creative geniuses which can now be seen to be of vital and continuing importance, and the occasional pursuit of a blind alley of development with all the heartbreak that entailed.

The names of the earliest pioneers are now almost forgotten. One was an Italian priest, Abbe Caselli, who in 1862 achieved sufficient success in transmitting handwritten messages and drawings over telegraph lines to attract the interest of Napoleon. And in 1881 an Englishman named Bidwell demonstrated an "electric distant vision" apparatus in which the picture to be transmitted was analyzed by a selenium light-sensitive cell moved up and down across the picture by a cam.

In 1908 Bidwell wrote a letter to the scientific journal *Nature* proposing for picture transmission the use of 1000 photoelectric cells, each linked to a separate light source in the receiver. Clearly a system requiring 1000 wires was not practical and means were required for conveying all of the picture information on a single wire. This had been realized long before Bidwell's 1908 letter and had given rise to the concept of "scanning," i.e., examination of the picture by a moving point in a series of closely spaced lines.

PICTURE ANALYSIS BY MECHANICAL SCANNING

Paul Nipkow

A little-known German inventor, Paul Nipkow, is credited with the invention in 1884 of the principle of mechanical scanning based on a rotating disk with a spiral of holes. As each hole moved across the picture a line was scanned, the successive line scans falling one below the other until the whole picture had been traced. It was Nipkow's mechanical scanning system that was resurrected more than 40 years later by John Logie Baird and used for a short time by the BBC in the world's first public television broadcasting system in 1936.

THE BIRTH OF ELECTRONIC TELEVISION: PICTURE ANALYSIS BY ELECTRONIC SCANNING

A. Campbell-Swinton (d. 1930)

Scottish electrical engineer A. Campbell-Swinton patented in 1911 an electronic-scanning picture transmitting and receiving system based on cathode-ray tubes which has been described as "an amazing piece of scientific clairvoyance, comparable perhaps to Charles Babbage's anticipation of the principle of the computer."[1] It covered the basic principles on which modern television cameras and receivers work.

Campbell-Swinton's early career had included activities as diverse as X-ray photography and introducing the young Marconi to Sir William

PAUL NIPKOW in BERLIN.

Elektrisches Teleskop.

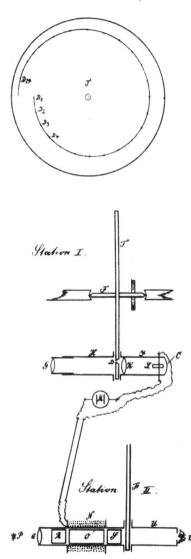

Figure 9.1. Paul Nipkow's mechanical scanning disk (1884).

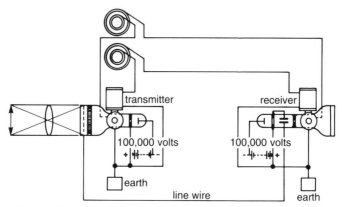

Figure 9.2. Campbell-Swinton's electronic television system (1911).

Preece of the British Post Office. He was also familiar with von Braun's work in Germany on the use of the newly invented cathode-ray tube to display the waveform of alternating currents.

He had noted Bidwell's 1908 letter in *Nature* proposing the use of 1000 photoelectric cells and 1000 wires as a picture transmitting system and looked for means whereby the whole picture information could be transmitted over a single wire. His 1911 patent disclosed a camera tube in which a cathode-ray beam of electrons scanned a mosaic of photocells, line by line.The scanning beam in effect switched on each photocell in turn, releasing the charge that had been accumulated as a result of the light falling on that cell for the whole time since the previous scan—a prerequisite for high sensitivity. A similar and synchronized scanning beam at the cathode-ray tube receiver generated the reproduced picture.

Campbell-Swinton's ideas for a wholly electronic television system were ahead of their time, as he himself realized; he modestly wrote:

> It is an idea only, and the apparatus has never been constructed. Furthermore it could not be got to work without a great deal of experiment and probably much modification.

A major step toward the practical realization of his proposal, which he described in his presidential address to the Röntgen Society in 1911, was ultimately achieved by Vladimir Zworykin in his 1923 "iconoscope" patent (see next page).

Campbell-Swinton was not alone in visualizing a cathode-ray tube receiver; he had been anticipated by Professor Boris Rosing at the Technological Institute in Petrograd. However, Rosing's experimental equipment used mechanical scanning for the transmitter and, for lack of suitable amplifiers for the weak photocell signals, failed to achieve other than a faint image.

Vladimir Zworykin

One of Rosing's students was the young physicist Vladimir Zworykin, who had also studied under the famous physicist Paul Langevin in Paris. During the First World War he served in the Russian Army Signal Corps and later joined the Russian Wireless Telegraph and Tele-

Figure 9.3. Dr. Vladimir Zworykin and the iconoscope.

phone Company and began to think about the problems of "seeing at a distance."

Perhaps inspired by his contact with Rosing, he came to the conclusion, independently of Campbell-Swinton, that electronic scanning—in which the virtually inertialess electron beam of a cathode-ray tube replaced the necessarily speed-limited rotating disk or mirror-drum of mechanical scanning—was essential for high-definition television. Furthermore it offered the prospect of continuing improvements in definition as the related technology improved.

Zworykin left Russia in 1919 after the Revolution to join the Westinghouse Electric and Manufacturing Company in Pittsburgh, Pennsylvania. The firm was not, however, receptive of his ideas; in spite of this he continued working on electronic scanning in his spare time and in 1923 filed a patent for his "iconoscope" which embodied the all-important concept of electric charge storage between scans. A rival electronic scanning system developed by P. Farnsworth in the United States, described by him as an "image dissector," did not embody the principle of energy storage with its sensitivity advantage.

Zworykin joined RCA and led their television research, giving a demonstration of electronic scanning to the Institute of Radio Engineers in New York in 1931, and a paper on this subject to the Institution of Electrical Engineers, London, in 1933. In England the firm of Electric and Musical Industries Ltd., which had links with RCA, began its own independent development of a camera tube—the "Emitron"—based on iconoscope principles. It was this camera tube that enabled the BBC to produce pictures of outstanding quality from the very beginning of its public television broadcasting service in 1936.[2]

Electronic scanning became important, not only for television broadcasting, but also for the cathode-ray picture tubes used in visual display units (VDUs) embodied in most computer and information access (viewdata/teletext) systems.

A LOST BATTLE FOR TELEVISION BROADCASTING

John Logie Baird (d. 1946)

John Logie Baird was a largely self-taught, highly prolific inventive genius who pursued with immense drive and confidence an approach

Figure 9.4. John Logie Baird.

to the design of a television broadcasting system that was nevertheless doomed to ultimate failure through its reliance on mechanical scanning principles. His efforts, however, had great value; above all they created an early public interest in, and awareness of, the possibilities of television as a public service, and undoubtedly spurred on a sometimes reluctant BBC to take television seriously.

Baird was the son of an impoverished Scottish minister of religion and suffered from ill health for most of his working life. His chief hobbies as a young man were experimenting with electricity and photography. Although he began studies at Glasgow University these were interrupted by the outbreak of the First World War. In the postwar years he found great difficulty in getting a job in England and went to the West Indies. There he set up small factories manufacturing soap and jam, but his businesses failed and he returned to England in 1922 broken in health and spirit. His biographer has written: "it was an unimpressive setting for one of the greatest developments in the history of invention."

Realizing he would never be a success as a trader, Baird turned his thoughts to an earlier dream derived from his hobbies of electricity and photography—the transmission of pictures over wires. He knew that photocells would generate an electric current when excited by light, and that the newly invented thermionic valves (see Chapter 5) could amplify weak currents to a useful level. Furthermore he was aware of the Nipkow disk with its spiral of holes as a possible means for scanning a picture.

In 1925 Baird set up, in an attic on Frith Street, Soho, London, a 30-line scanning disk camera and receiver with a neon tube as a light source. For a picture source he used the head of a tailor's dummy—and at one stage a very scared and reluctant boy from a neighboring office! With this improvised equipment, made from scrap material, he was able to transmit crude pictures, small in size and not much more than black silhouettes against a red background, from one side of his laboratory to the other.

Clearly much more needed to be done, better technical resources and capital were needed. He approached the Marconi Company but since his apparatus had no patent protection, because of its dependence on the Nipkow disk, the Company was not interested. Partly to stimulate public interest and attract capital, Baird embarked on a series of demonstrations and patents involving the televising of subjects in darkness by infrared rays, color television using a triple-spiral Nipkow disk, stereoscopic television, and even the recording of picture signals on a gramophone disk. Needless to say, with the limited technology of the time, these were of crude picture quality.

Backers were eventually found, attracted no doubt by Baird's demonstrated inventiveness and drive, and the Baird Television Company was formed. By 1926 a 30-line Baird "televisor" had been put on the market, and kits of parts were also available for home assembly.

At this stage Baird knew he had to persuade the BBC to transmit his 30-line television signals on their medium-wavelength sound broadcast transmitters. At first the BBC was not interested, understandably in view of the poor picture quality. Eventually, after some persuasion by the then Post-Master General Mr. Lees-Smith, they agreed that regular but short public television transmissions could start from London in September, 1929. This was a great day for Baird and a series of broadcasts followed, including the finish of the Derby in 1931. By this time he had replaced his spinning disk camera by a rotating mirror-drum giving some improvement of picture quality, but limited by 30-line definition.

While EMI and the Marconi Company were pursuing the development of a high-definition television system based on electronic scanning and the Emitron camera, the Baird Television Company struggled on to improve the definition of their mirror-drum camera, first to 120 lines and then to 240 lines.

Figure 9.5. Baird's 1926 "televisor." The black-and-red picture was barely 1 inch in size and had to be viewed through a magnifying lens.

THE WORLD'S FIRST HIGH-DEFINITION PUBLIC TELEVISION BROADCAST SERVICE IS BORN

A marriage of the camera skills that had been developed by EMI with the VHF transmitter and aerial design know-how in the Marconi Company was formally agreed in March of 1934 with the formation of the Marconi–EMI Television Company.[3] The timing was fortunate since the trial period for the Baird 30-line system had come to an end and the Government had set up a committee under Lord Selsdon to advise the Post-Master General on the future of television broadcasting in the United Kingdom. When in 1935 the Selsdon Committee produced its report, it recommended a high-definition system with not less than 240 lines per picture frame. Of the possible contenders this left only Marconi–EMI and the Baird Television Company; to resolve the problem of selection it was decided that each should provide service from Alexandra Palace, London, on alternate weeks so that a fair evaluation of each could be made.

The Baird Company had pinned its faith on mechanical scanning and shown great ingenuity in achieving a 240-line standard; it used a "flying-spot" scanner for normal studio work and an intermediate film process for large studios and outside broadcasts.

Marconi–EMI, armed with the electronic scanning Emitron camera, elected with great courage and foresight to go to a higher-definition 405-line standard. This choice was largely the result of the brilliant research team at EMI led by Isaac Schoenberg, and which included Alan Blumlein (whose contribution of stereophony to sound broadcasting was discussed in Chapter 8).[4,5]

THE EVOLUTION OF THE 405-LINE TELEVISION STANDARD WAVEFORM

Alan Blumlein

A satisfactory television signal waveform has to include features that readily enable frame and line synchronization, and alternate field interlacing, to be achieved in receivers; the line frequency should desirably be achieved by frequency multiplication from the mains frequency of 50 Hz. In particular the lock to the mains frequency helped to avoid moving "hum" bars on the reproduced picture at the receiver.

It is said that the manifestly successful 405-line waveform finally adopted as the British standard was created one Sunday on a breadboard at Blumlein's home, with the help of his colleagues Cork and White. The "405" came from four stages of 3-times frequency multiplication, followed by 5-times multiplication, of the 50-Hz mains frequency.

The basic 405-line waveform evolved by Blumlein and his colleagues served television development very well and stood the test of time; with appropriate modifications it later evolved into the 625-line waveform and proved capable of accommodating color and additional services such as Ceefax/Oracle teletext.

THE BAIRD/MARCONI–EMI TELEVISION SYSTEM TRIALS AT ALEXANDRA PALACE

Each company provided its own vision and sound transmitters, working into a common aerial designed by C. S. Franklin of the Marconi Company. The vision transmitters operated at 45 MHz with a 4-MHz bandwidth, the sound at 41.5 MHz.

The studio equipment was in striking contrast: the EMI camera

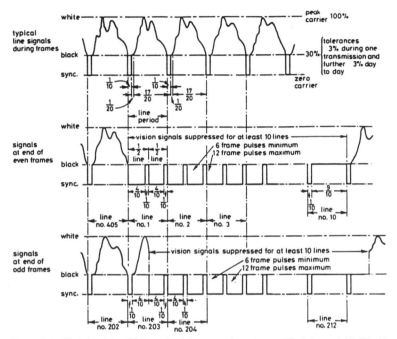

Figure 9.6.   The Blumlein 405-line television waveform (as used by Marconi–EMI Ltd.).

was compact and mobile, whereas the Baird flying-spot scanner, and especially the intermediate film unit, was relatively immobile.

Experimental public transmissions commenced in October, 1936, and the service was given a formal opening in November, 1936.

The rival systems were used alternately for a time but it rapidly became apparent that the Baird flying-spot scanner, which required a studio in semidarkness and was troublesome to the performers, was at a disadvantage compared with the Marconi–EMI image-orthicon system. (The image-orthicon camera was a development from the iconoscope using electron-beam scanning and charge storage that improved both picture quality and sensitivity.[9]) An even greater disadvantage of the intermediate film process was the immobility of the equipment, which made it unsuited to outside broadcasts.

In February, 1937, the Selsdon Committee recommended that the Marconi–EMI system—notably on the grounds of its greater flexibility and scope for further development—carry the permanent program service.

It is natural to sympathize with John Logie Baird who labored unceasingly for many years, often in poverty and ill health, but it became clear that the mechanical scanning system had reached a peak of development and had no future.

An excellent and detailed history of the development of television in Britain during the formative years 1923–1939 has been given by Professor R. W. Burns.[7]

THE TELEVISION SERVICE GROWS, AND THE PICTURE IMPROVES

From its beginning in 1936, and apart from a gap in the war years, the 405-line monochrome BBC television service expanded throughout the United Kingdom, using the VHF frequency bands 41–68 and 174–216 MHz. In 1953 the commercial Independent Television service was inaugurated, and by 1964 some 97% of the population could receive both programs.

With the advent of a third service (BBC 2) in 1964, a 625-line standard was adopted and a move made to use the UHF band 470–960 MHz. The 625-line standard gave a notable improvement in picture quality, while the use of UHF enabled receiving-aerial directivity to be sharpened compared with VHF, thereby minimizing interference from unwanted transmissions and reducing radio-wave echoes from buildings, etc.

The technical characteristics of the 625-line standard were the subject of detailed study by the General Post Office, which had a responsibility through Parliament and the Post-Master General for the regulation of broadcasting (see Refs. 4 and 5), the BBC and the ITA, and the U.K. television industry. They were adopted internationally by the International Radio Committee of the International Telecommunication Union and used throughout most of Europe. Advances were also made in the design of camera tubes—notably the image orthicon[9]—and picture tubes for receivers, that further improved picture quality. These technical advances were perhaps not especially innovative in themselves, they were more the result of step-by-step logical development rather than highly creative new concepts. But in the latter category must come the introduction of color to television—aptly described by G. H. Brown[6] as "Searching for the Rainbow."

COLOR COMES TO TELEVISION

The creation of a high-quality viable color television system, which required a remarkable blend of sophisticated scientific study, detailed engineering design, careful economic evaluation, and ultimately international agreement, may be regarded as one of the crowning achievements of 20th-century technology, in which the United States, the United Kingdom, France, and Germany were major participants.

As early as 1928 John Logie Baird had demonstrated a color system based on a Nipkow scanning disk with three sets of spirals; in one the holes were covered by a red filter, a second by a green filter, and a third by a blue filter. However, by 1942 Baird and an assistant, E. Anderson, were experimenting with a two-color cathode-ray picture tube they called a "telechrome"[10]—a development that ended with Baird's death in 1946 but which showed a vision beyond electromechanical systems.

Over the years 1940 to 1953 there was intense activity and competition in the laboratories of commercial organizations in the United States to develop a satisfactory color television system; the prize was immense, a new mass market for tens of millions of television receivers, and a massive stimulus to the advertising industry by showing its products in glowing color.

There were two main contenders for the prize:

- The Columbia Broadcasting System and its engineer Peter Goldmark with a "field sequential" system
- RCA and George H. Brown with a "dot sequential" system.

The field sequential system bore a strong resemblance to Baird's 1928 proposal in that complete red, green, and blue pictures (fields) were transmitted one after the other. Although it had the advantage of simplicity, the field sequential system required three times the bandwidth of a black-and-white picture offering the same definition. If the bandwidth was reduced by lowering the field rate, problems of flicker arose.

RCA's proposals aimed at achieving compatibility with existing black-and-white receivers and fitting color into the same video bandwidth (4 MHz in the U.S. 525-line television standard). This created technical problems of substantial difficulty, needing great ingenuity, skill, and determination for their solution.

The choice of a national color television standard rested with the U.S. Federal Communications Commission (FCC) which held a series of hearings on the various proposals between 1946 and 1953. However, some of the Commissioners do not appear to have understood all of the subtle technicalities in the presentations made to them. And these difficulties were compounded by the intense commercial rivalry between the participants, and the feverish pace at which modifications were made to the proposed standards to overcome objections from competitors. The story of this period and the infighting that occurred has been told in illuminating detail by George Brown in his book "*And Part of Which I Was.*"[6]

Certain highly innovative ideas, which can now be seen as crucial, emerged during this period of intense development.

Alda V. Bedford: The "Mixed Highs"

The crowding of color television into a radio channel bandwidth of 12 MHz was made possible by the principle of "mixed highs," developed by Bedford and adopted by RCA. It recognized that the eye fails to distinguish color in the fine detail of a picture, and made it possible to transmit limited-bandwidth red and blue information on a "chrominance" subcarrier together with a full-bandwidth green "luminance" channel. In 1954 Bedford was presented with the Vladimir Zworykin Award by the Institute of Radio Engineers for "his contribution to the principle of mixed highs and its application to colour television."

Alfred C. Schroeder: "The Color Dot Triad"; Norman Fyler and W. E. Rowe: The Shadow Mask Tube

Earlier demonstrations by RCA to the FCC had involved the use of separate red, green, and blue picture tubes and the superposition of these images by dichroic mirrors—an arrangement that had registration problems and was too bulky and costly for a mass-produced domestic television receiver.

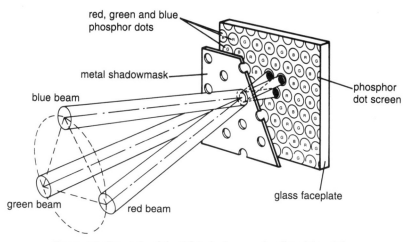

Figure 9.7. Principle of the RCA shadow-mask color picture tube.

A. C. Schroeder had been working with Bedford and Kell of RCA on the problems of color television and developed the idea of a single color tube with groups of three red, green, and blue light-emitting phosphor dots deposited directly on the inside end face of the tube. Three beams of electrons were to be aimed at the phosphors through holes in a metallic grid, arranged so that one beam struck a green dot, another a red dot, and a third a blue dot. Schroeder filed a patent for this idea in 1947.

Fyler and Rowe developed the idea further by applying the phosphor dots directly on the inside of the end of the tube through the shadow mask, thereby automatically ensuring correct registration.

Although many variants of the shadow mask tube have since been devised, the effectiveness of the original design and its manifest suitability for manufacture in quantity undoubtedly made a major contribution to the success of color television in the United States and Europe.

THE U.S. NATIONAL TELEVISION SYSTEM COMMITTEE

At the request of industry the FCC set up in 1950 a National Television System Committee (NTSC) with wide representation and powers

to investigate proposals and make representations to the FCC on technical standards. In this it was assisted by the Hazeltine Laboratories which
investigated the CBS, RCA, and other proposals, but which took no
part itself in the FCC hearings. After a detailed and massive program
of study and demonstrations to optimize the color system performance
and secure the best possible use of the radio-frequency spectrum, the
NTSC proposals were adopted by the FCC in December, 1953.

According to G. Brown,[6] the major contributions came from RCA,
notably the use of a subcarrier where hue was determined by phase
and saturation by amplitude, the signal burst to synchronize color,
and the principle of "mixed highs." The Hazeltine Laboratories were
credited with the constant luminance principle and important contributions to the composition of the color subcarrier. But all in all it was an
outstanding example of cooperation on the part of a large number of
engineers in a joint effort to achieve the best solution to a major technical
problem of common concern.

By 1979 more than 50 million NTSC-standard color television sets,
valued at $25 billion, were in use in the United States.

COLOR TELEVISION COMES TO BRITAIN AND EUROPE

Soon after the end of World War II, television engineers in the
United Kingdom and Europe began following developments in color
television in the United States, and began their own investigations.

The British Post Office, with its responsibility for the regulation of
broadcasting, began a detailed investigation of color systems, including
studies to define the technical parameters for optimum picture quality
and efficient use of the radio spectrum.[5] Similar studies were carried
out by the BBC and the television industry, notably by Marconi–EMI.[4]

In France the Compagnie Française de Television was working on
a system that sought to minimize the effects of transmission delay
distortion, e.g., on some cable and microwave links, by transmitting the
color information on alternate pairs of picture lines in sequence with a
delay device to combine the two, hence the name SECAM (système
électronique couleur avec mémoire). The system also used frequency
modulation of the subcarrier instead of the suppressed-carrier amplitude modulation in the NTSC system.

In Germany the firm of Telefunken developed a system based on alternating the color phase line by line, hence the name PAL (phase alternation line), invented by Walter Bruch.

The problem of deciding on an international standard for color television broadcasting was one for the International Radio Consultative Committee (CCIR), a body of the International Telecommunication Union in Geneva, to resolve. Beginning in 1955 the CCIR Study Group XI held a series of meetings and witnessed demonstrations of color systems in the United States, the United Kingdom, France, and Germany. A plenary session of the CCIR was convened in Oslo, Norway, in 1966 with the aim of reaching an international agreement. However, by that time the various national positions had become firmly entrenched and political as well as commercial rivalries determined the outcome. The United States and Japan decided to remain with the 525-line NTSC system, the United Kingdom and Western Europe (except France) opted for the 625-line PAL, and France and the USSR chose SECAM.

The BBC commenced its color transmissions on the PAL system in 1967, as did the Independent Television Authority. Fortunately, program interchange between the various systems and standards conversion later became relatively straightforward with the invention of digital techniques and the microcircuit (see Chapters 11 and 12).

George Brown, a vigorous defender of the American NTSC system, gives a forthright view of the behind-the-scenes activities that led up to the CCIR meeting in Oslo, and a critical commentary on the difficulties of arriving at decisions by committee, in his book.[6]

THE RECORDING OF VIDEO SIGNALS

The early recorded television programs were based on the use of "flying-spot" telecine equipment, in which a moving photographic film was scanned by a spot of light focused on it from a cathode-ray tube. Telerecordings were also made by filming the picture on a high-quality television monitor. Since they involved the use of photographic film with its processing problems, these were not highly convenient methods of recording.

Magnetic tape recording of audio signals had been made in Germany during the war, but the recording of a television signal was much more difficult since it involved frequencies several hundred times higher than for audio. An early attempt at recording video signals on magnetic tape was made by the BBC Research Laboratories at Kingswood Warren, Sussex, but required excessive tape velocities and this limited recording time.

The first commercially successful magnetic tape video recording system was developed in the United States by the Ampex Corporation and came into use in 1956. It involved the use of a moving magnetic tape, 2 inches wide, with a rapidly spinning drum carrying four recording heads moving across the tape instead of along it. By arranging that the spinning drum laid down transverse tracks slightly inclined to the length of the tape, each short length of the tape recorded a complete scan of one picture frame. The first uses of magnetic tape video recording were in television studios for recording and reproducing complete television programs of an hour or more. However, it became apparent that, provided problems of size, ease of use, and cost could be solved, there was an immense market for video recording equipment for use in the home.

The challenge to produce video recording equipment for the domestic market was first taken up by Philips in Holland who demonstrated in 1972 a domestic recorder using 0.5-inch magnetic tape in a compact cassette. This had the useful capability for wipeout and reuse of the magnetic tape, facilitating the recording of broadcast television programs "off-air." By the mid-1970s Sony and JVC in Japan had entered the competition and produced their (incompatible) Betamax and VHS systems—in the event the VHS system captured the bulk of the domestic market.[8]

The availability of prerecorded videotape cassettes, providing up to 2 or 3 hours of television entertainment, has proved very attractive to domestic viewers.

Perhaps looking back to John Logie Baird's idea of recording still pictures on a gramophone record, Decca, Telefunken, and Philips in Europe, Matsushita and JVC in Japan, and RCA in the United States began in the mid-1970s to explore the possibility of recording video signals by microscopic mechanical impressions on cylindrical or flat circular disk plastic records—the objective being low-cost manufacture in quantity by pressing the recorded video disks, analogous to the manufacture of gramophone audio disk records.

LaserVision, a successful video disk technology, has been created by Philips, Holland using a digitally modulated laser beam to impress minute pits on a helically scanned flat circular disk in the recording mode, another laser beam being used to scan and detect the pits in the reproducing mode. However, unlike magnetic tape recording, the mechanically impressed video disk cannot be wiped clean and reused.[8]

None of the currently available video recording technologies is well adapted to the multiple-access video library concept of the future; this demands a solid-state (micro- or optotechnology) device approach (see Chapter 20).

HIGHER-DEFINITION AND SATELLITE DIRECT BROADCASTING TELEVISION

Cable television local distribution networks and direct broadcasting from satellites to viewers' homes avoid the bandwidth restrictions imposed by the limited space available in the UHF radio spectrum for conventional television broadcasting. There is thus the prospect of improving picture definition, e.g., by increasing the number of lines per frame above the 625-line standard, enabling larger screens to be used, both in the home and for public viewing.

One such approach, known as HDTV (high-definition television), pioneered by the Japanese Broadcasting Authority NHK and Sony proposes a totally new standard using 1125 lines at 30 frames per second, requiring completely new designs of receivers.

Another approach, aimed particularly at satellite transmission, is the MAC (Multiplexed Analogue Components) system devised by the Independent Broadcasting Authority in the United Kingdom and which seeks an evolutionary development from the existing 625-line, 25 frames per second standard. The MAC system minimizes the effect of the higher levels of noise at the upper end of the video spectrum characteristic of transmission by satellite, where the color information is normally transmitted, by sending each line's color information immediately before its monochrome component, the two being time compressed to fit within the available time. At the receiver the two components are time stretched back to their normal lengths and reunited. The time compression increases the transmitted signal bandwidth; nevertheless, the overall effect is to enable the domestic receiving aerial to be smaller than would otherwise be required.[9] The European Community, in a bid to bypass Japanese dominance in the consumer electronics field, sponsored in 1986 the development of an HD–MAC television system.

However, analogue methods of television signal transmission over coaxial cables and microwave links are being replaced by digital methods which offer high quality and improved flexibility. As telecommunication networks move into digital methods for the transmission and switching of signals—exemplified by the Integrated Services Digital Network (Chapter 19)—with integration of telephone, data, facsimile, and teleconferencing services in a common network, so the economic and operational pressures to include broadcast-type television in such a network are growing. The inclusion of television signals in the digital network is made possible by sophisticated video signal bit-rate reduction techniques that reduce the bit rate for transmission by a ratio as high as 100 to 1 without significant impairment of picture quality.

Present-day HDTV standards envisage a 9 to 16 ratio of picture format with 1250 lines, compared with the current widely used 3 to 4 format with 625 lines, the more rectangular aperture making better use of picture information and offering one better suited to large-screen presentation.

The problems of achieving compatibility between HDTV and existing terrestrial 625/25 TV broadcasting and transmission networks appear substantial; with satellite TV broadcasting the constraints are fewer and this would seem to offer the best prospect for early introduction of a domestic HDTV service.

But it may well be that digital techniques will progress into the UHF radio broadcasting band, which could then be replanned to provide many more TV channels, together with radiotelephone, facsimile, and similar services for mobile users.

It is clear that there is scope for continuing innovation in television broadcasting: in addition to the ongoing quest for improved picture quality, a viable three-dimensional television display appears so far to have eluded the innovators.

REFERENCES

1. *Eureka: An Illustrated History of Invention*, edited by Edward de Bono (Thames and Hudson Ltd., Cambridge, 1974).
2. Keith Geddes and Gordon Bussey, ''Television: The First Fifty Years,'' National Museum of Photography, Film, and Television, Trustees of the Science Museum, 1986; *John Logie Baird: Sermons, Soap and Television* (Royal Television Society, London, 1988).
3. W. J. Baker, *A History of the Marconi Company* (Methuen, London, 1970).
4. *The History of Television: From Early Days to the Present*, Institution of Electrical Engineers Conference Publication No. 271, November 1986.
5. W. J. Bray, ''Post Office Contributions to the Early History of Television,'' Ref. 4, p. 136.
6. G. H. Brown, *And Part of Which I Was: Recollections of a Research Engineer* (Angus Cupar Publishers, Princeton, NJ, 1979).
7. R. W. Burns, *British Television: The Formative Years (1923–1939)*, Institution of Electrical Engineers History of Technology Series Vol. 7 (Peter Peregrinus, London, 1986).
8. David Owen and Mark Dunton, *The Complete Handbook of Video* (Penguin Books, Baltimore, 1982).
9. A. Rose, P. K. Weimer, and H. B. Law, ''The Image-Orthicon—A Sensitive Television Pickup Tube,'' *Proc. IRE, 34*, No. 7, July 1946.
10. R. Herbert, ''Guns for Peace (Baird's 'Telechrome' Colour Cathode-Ray Picture Tube),'' *IEE Rev.*, July 1994.

10

The Engineers of the Early Multichannel Telephony Coaxial Cable Systems

The First Transatlantic Telephone Cable

THE GROWTH OF INTERCITY TELEPHONY

Beginning in the 1920s the need for ever more telephone circuits between cities in the United States and in Europe was at first met by carrier telephone systems providing some tens of telephone circuits on pole-mounted open wires or on multipair wire cables, using the frequency-division multiplex principles outlined in Chapter 6. However, as the numbers of circuits on each pair grew, problems were encountered from crosstalk between pairs and the increasing attenuation of the signals. As the demand grew into hundreds and more circuits on each intercity route, a more efficient and cost-effective solution was sought.

The engineers of the American Telephone and Telegraph Company must be given the major credit for evolving the "frequency-division multiplex (FDM) analogue coaxial cable system'' which, together with FDM microwave radio-relay systems (Chapter 11), became the dominant modes of intercity transmission from the 1940s to the 1980s.[1] From the 1980s time-division multiplex (TDM) digital systems (Chapter 13) began to supplement analogue FDM systems and eventually displace them.

The most important innovative ideas and concepts that created the technological foundation for the FDM coaxial cable system are outlined below.

THE COAXIAL CABLE

Theoretical studies of a transmission line embodying cylindrical inner and outer conductors, i.e., a coaxial cable, were carried out by several Victorian scientists including Kelvin, Heaviside, Rayleigh, and J. Thomson, and in 1909 A. Russell published in the *Philosophical Magazine* a detailed analysis that enabled the impedance and loss to be calculated as functions of frequency. The first practical use of coaxial cable was in submarine telegraph systems (Chapter 3). Early applications were also found in the feeder systems of shortwave radio aerials, notably by C. S. Franklin of the Marconi Company (Chapter 7). Franklin derived, in 1928, the optimum ratio (3.6) of outer to inner conductor diameter for minimum loss. The relatively low attenuation in coaxial cables, even at high frequencies, made them suitable also for television feeders operating at VHF and UHF.

From many points of view the coaxial cable was ideally suited for a long-distance multichannel telephony transmission system—notably the freedom from crosstalk to and from neighboring cables, the built-in screening from other forms of electromagnetic interference such as power lines, and attenuation/frequency characteristics that offered a wide useful bandwidth capable of accommodating large numbers of telephone channels.

H. A. Affel and L. Espenscheid

Recognizing these favorable characteristics, H. A. Affel and L. Espenscheid of AT&T/Bell Laboratories filed a patent for a coaxial cable telephony system in May, 1929—a patent that set the stage for one of the most important developments in telecommunications.

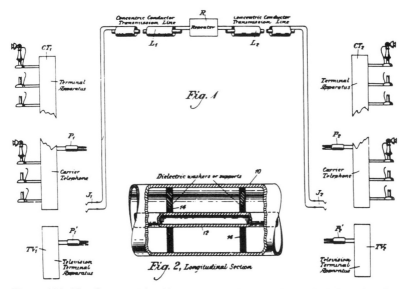

Figure 10.1. The first coaxial cable carrier telephone system patent, issued to L. Espenscheid and H. A. Affel, Bell Laboratories, May, 1929.

TAMING THE CABLE LOSS AND NOISE

S. A. Schelkunoff and O. J. Zobel

The earlier theoretical studies of coaxial cable loss/frequency characteristics made by the Victorian scientists and mathematicians and others were extended and refined by S. A. Schelkunoff of the Bell Laboratories on the basis of electromagnetic field theory to predict the effect of departures from an idealized geometry such as might occur in practice, and to allow for losses in insulators supporting the inner conductor. Broadly, the "ohmic" loss related to the skin effect of the inner and outer conductors was found to increase as the square root of the frequency, with an additional loss related to the insulators appearing in proportion to the frequency when the latter was sufficiently high. The problem of equalizing the variation of cable loss with frequency by suitably designed networks was solved by the work of O. J. Zobel[2] and others at the Bell Laboratories. There remained the need to compensate for the overall loss of the equalized cable by valve amplifiers or repeaters;

however, there were limits to the usable amplification set by valve and cable noise, and this in turn determined the maximum permissible spacing between repeaters.

J. B. Johnson and H. Nyquist

A classic study of noise in valve amplifiers related to the random nature of electron emission had been made by W. Schottky of Siemens and Halske in Germany in 1918.

However, when valve noise had been reduced to a sufficiently low level, J. B. Johnson of Western Electric was able to demonstrate in 1925 that a measurable noise contribution came from the circuit attached to the valve input. This noise was identified with thermal agitation of the electrons in the input circuit.[3] It was H. Nyquist who analyzed the effect mathematically from thermodynamic considerations and revealed the classic formula for the noise power P in a bandwidth B as

$$P \text{ (noise)} = 4 \times KTB \quad \text{(watts)}$$

where K is Boltzmann's constant and T is the absolute temperature.[4]

With the problems of cable loss, equalization, and noise under control, it remained to achieve stable, linear amplification of the FDM multichannel telephony signal. Amplifier gain stability was vitally important, especially in long submarine cable FDM systems where an overall loss of perhaps 1000 decibels (10 multiplied by itself 100 times) had to be stabilized to within a decibel (1.26 times), year in and year out. Linearity, that is, a precise proportionality between the amplitudes of the output and input signal levels up to a defined "overload" point, is essential in order to avoid crosstalk between the telephony channels. However, the input/output characteristics of the valve amplifiers used in repeaters lacked the necessary linearity to meet the stringent requirements for high-quality, crosstalk-free transmission, especially in systems with many repeaters in tandem. (An additional requirement—phase linearity—later became evident when coaxial cable systems were used for the transmission of television signals, in order to avoid waveform distortion.)

The solution to these problems came from an invention of remarkable simplicity and power—"negative feedback"—conceived of by H. S. Black of the Bell Laboratories, a concept that was to have repercus-

Figure 10.2. H. S. Black and a 1930 negative feedback repeater.

sions far outside telecommunications and even on economic, political, and philosophical theory.

NEGATIVE FEEDBACK—A KEY INVENTION OF THE 20TH CENTURY

H. S. Black

M. J. Kelly, President of Bell Laboratories, said on the occasion of the presentation of the American Institute of Electrical Engineers Lamme Gold Medal to Black in 1957:

Although many of Harold Black's inventions have made great impact, that of the negative feedback amplifier is indeed the most outstanding. It easily ranks with Lee De Forest's invention of the audion as one of the two inventions of broadest scope and significance in electronics and communications of the past 50 years. Without the stable, distortionless amplification achieved through Black's invention, modern multi-channel trans-continental and trans-oceanic communication systems would have been impossible.

One of Black's first attempts to solve the problem in the 1920s envisaged an amplifier in which the output was reduced to the same amplitude as the input; by subtracting one from the other, only the distortion products remained, and these could then be amplified in a separate amplifier and used to cancel the distortion products in the original output, i.e., a "feed-forward amplifier." Although this invention later found application in the high-power amplifiers of single-sideband shortwave radio transmitters, the need for precise balancing over a wide frequency band made it unsuitable for FDM multichannel telephony systems.

Black continued to wrestle with the problem for many years, without success. How it was solved is well told in Ref. 5:

Finally, on 2 August 1927, while on the Lackawana Ferry crossing the Hudson River on the way to work, he had a flash of insight, famous in the annals of Bell Laboratories. He realized that by employing "negative feedback", that is by inserting part of the output signal into the input in reversed phase, virtually any desired reduction in distortion could be obtained by a sacrifice in amplification. He sketched the diagram and scribbled the basic equation on a page of the New York Times he was carrying. He had it witnessed upon arrival at his office: negative feedback had been invented.

The key element in the invention is a passive feedback network, the loss/frequency characteristic of which determines the overall performance of the repeater, making it substantially immune to changes of valve amplification. By suitably proportioning the gain of the amplifier and the characteristics of the network, nonlinear distortion can be reduced to any desired degree.

However, the path to practical realization was long and complex. It included a long and drawn-out battle with the U.S. Patent Office to cover the many possible applications of negative feedback; success was ultimately reached with the granting of a patent in December, 1937.

Much more subtle and requiring great ingenuity, mathematical skill, and clear realization of the physical phenomena involved was

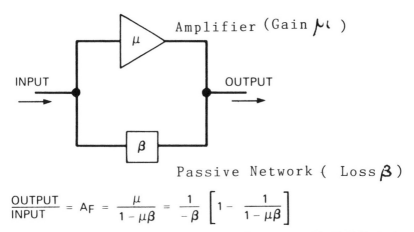

$$\frac{\text{OUTPUT}}{\text{INPUT}} = A_F = \frac{\mu}{1 - \mu\beta} = \frac{1}{-\beta}\left[1 - \frac{1}{1 - \mu\beta}\right]$$

Figure 10.3. The key invention of negative feedback, as conceived by H. S. Black of Bell Laboratories, August, 1927.

the evolution of design criteria for the passive network that enabled oscillation around the feedback loop to be avoided. Much credit for the determination of viable design criteria must go to H. Nyquist and H. W. Bode of the Bell Laboratories.[6,7]

THE POWER OF A GREAT IDEA

The applications of Black's invention and the Nyquist/Bode design criteria were not limited to repeaters for telecommunication systems: they apply to a wide range of acoustic, mechanical, and electrical systems, and especially to control systems in which dynamic stability is of vital importance, e.g., automatic steering systems in aircraft and rockets. It is to the credit of the Bell Laboratories that the theory and design techniques evolved there have been made freely available to the electronics and control industries worldwide.

But the concept inherent in negative feedback has, at least potentially, much wider practical and philosophical applications. For example, in the human nervous system the control of eye or hand movement in carrying out a task is dynamically stabilized by feedback, and can fail if the feedback is absent or distorted. The relationships between government and the governed, and between production and demand in economic systems, require adequate feedback if stability and smooth progress are to be achieved. Perhaps our political and economic masters ought look more closely at Black's invention!

PACKING THE TELEPHONE CHANNELS TIGHTLY: THE ADVENT OF THE QUARTZ CRYSTAL BANDPASS FILTER

The economic utilization of the usable freqency band on a transmission medium such as a wire pair or coaxial cable requires the close packing of telephone channels in the frequency spectrum: the greater the number of channels, the lower is the cost per channel. For similar economic reasons the voice band of frequencies is customarily restricted to 200–3300 Hz, the minimum necessary for telephonic speech of good intelligibility and satisfactory recognizability. The spacing of channels on carrier systems has been standardized at 4 kHz by international agreement via the International Telephone and Telegraph Consultative Committee of the International Telecommunication Union, as has the arrangement of blocks of channels in the frequency spectrum. The standardization of these and other key parameters in FDM systems has played a vital role in establishing telecommunication links between countries throughout the world.

A key factor in the design of an efficient FDM system is a bandpass filter providing low-loss transmission of the voice band 200–3300 kHz when transposed to carrier frequencies, and the rejection of all other signals, including those in the adjacent 4-kHz-wide carrier channels, by some 60 decibels (10 multiplied by itself six times)—a very stringent requirement. While a solution to this problem is possible by conventional inductance/capacitance filter design using inductors with low-loss magnetic cores, a more compact and efficient design using quartz crystals as circuit elements was devised by engineers of the Bell Laboratories, the evolution of which has been described by O. E. Buckley.[8]

The very low damping (Q factor) of the piezoelectric quartz crystal enables a very sharp cutoff to be obtained at the transmission band edges, permitting efficient utilization of the 4-kHz channel spacing. It also facilitates the selection of a single sideband, and rejection of the

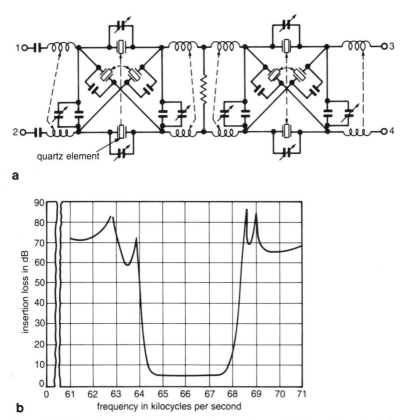

quartz element

a

b   frequency in kilocycles per second

Figure 10.4. The channel bandpass filter using quartz crystal elements (AT&T). (a) Circuit; (b) response.

unwanted sideband and the carrier, from an amplitude-modulated double-sideband signal. For convenience in design of the quartz crystal elements, a basic 12-channel group, using frequencies from 60 to 108 kHz, was agreed internationally via the CCITT. This forms a building block from which basic supergroups of 60 circuits and hypergroups of 960 circuits can be assembled forming, for example, coaxial cable systems bearing 2700 circuits in a 12-MHz band or 10,800 circuits in a 60-MHz band.[9]

COAXIAL CABLE SYSTEM PIONEERS
IN THE BRITISH POST OFFICE

Engineers of the British Post Office Engineering Department had maintained close contact with their counterparts in AT&T and the Bell Laboratories in the development of long- and shortwave radio systems from the 1920s to the 1940s, and it was natural that their attention should be drawn to the American work on the development of coaxial cable systems during the later part of that period. In the British Post Office the early exploratory studies of FDM telephony and television coaxial cable systems, leading to the design and manufacture of equipment for the first field trial systems in the 1930s, were carried out by a small group of engineers in the Radio Experimental Branch of the Engineering Department at the Post Office Research Station at Dollis Hill, NW London. In general, it fell to the radio engineers to carry out much of this work since their experience in the design of radio-frequency circuits and components matched the requirements of the new coaxial cable systems.

The engineers who carried much of the responsibility for pioneering coaxial cable system development in the British Post Office, under the direction of A. H. Mumford (later Engineer-in-Chief), were:

- Dr. R. A. Brockbank: Negative feedback FDM repeater design
- H. O. Stanesby: Quartz crystal filter design
- Capt. C. F. Booth: Quartz oscillator and filter element design and manufacture
- Dr. R. F. Jarvis: Coaxial cable and television repeater design
- H. T. Mitchell and T. Kilvington: Television applications

In particular, Dr. Brockbank and a colleague, C. Wass, made a classic study, much used in repeater design, of the loading of such amplifiers by multichannel FDM signals and the linearity requirements to avoid interchannel crosstalk.[10]

Their work gave a lead to British Telecommunications Industry and led to a fully engineered FDM telephony field trial on the London–Birmingham route in 1938, using 0.45-inch inner diameter coaxial pairs.[11] The cable comprised four such coaxial pairs, two for FDM telephony and two for television. The telephony pairs provided a total of 400 go

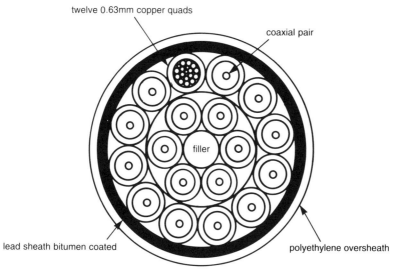

Figure 10.5. Eighteen-pair, 2.6/9.5-mm coaxial cable used in the British Post Office/ Telecom trunk network.

and return telephone circuits in a frequency band between 50 and 2100 kHz, with a channel spacing of 4 kHz. The television channel was limited to a bandwidth of 1.6 MHz, reasonably adequate for 405-line television signals but which was envisaged as needing upgrading in the future.

As noted earlier, the opening of the television broadcasting service from Alexandra Palace, London, in 1936 created a strong public demand for television broadcasting in other cities in the United Kingdom and a corresponding pressure to develop intercity television links. The outbreak of the Second World War in 1939 delayed further development of coaxial cable systems until the period following the end of the war in 1945 when a growing need for more intercity telephone circuits gave a new impetus to development.

As the years progressed from the 1940s toward the 1980s a growing family of coaxial cable FDM telephony systems was developed, with capacities of up to 10,800 circuits on each 2.6/9.5-mm coaxial pair, and also small-bore (1.2/4.4 mm) cables with up to 1920 circuits per pair.

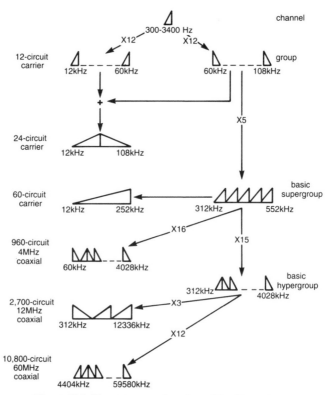

Figure 10.6. Frequency spectra of coaxial cable systems.

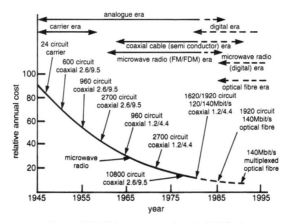

Figure 10.7. Relative annual costs (100 km).

With this progression has come a substantial decrease—of nearly ten to one—in the relative annual cost per telephone circuit, a remarkable tribute to the farsightedness of the pioneers who made it possible.

THE FIRST TRANSATLANTIC TELEPHONE CABLE 1956: A JOINT U.S., U.K., AND CANADA UNDERTAKING

AND WHEREAS it is desired to provide a submarine cable system between the United States and Canada on the west, and the United Kingdom in the east . . .

These words in a formal legal agreement between the British Post Office, the American Telegraph Co., and the Canadian Overseas Telecommunication Corporation marked the beginning of the most outstanding engineering achievement in telecommunications during the first half of the 20th century, nearly 100 years after the first transatlantic telegraph cable. The legal agreement, and the associated contract, represented an act of faith on the part of the designers and manufacturers, on both sides of the Atlantic, to create in the short space of 3 years a long multichannel telephony submarine cable system that demanded higher standards of precision in the design of the cable and repeaters, and longer life expectancy than had ever before been achieved. A detailed account by British and American engineers of the design, manufacture, and laying of the system is given in Ref. 12.

It was a stupendous pioneering undertaking. Some 4500 miles of coaxial cable had to be made to the most exacting specification ever devised, and new machinery had to be designed for laying the cable in waters up to 2.5 miles deep. Surveys of the transatlantic route had to be carried out to select the most suitable. One hundred forty-six repeaters had to be built to withstand the rigors of laying and the extreme water pressures in the deep ocean, and capable of functioning without attention for at least 20 years.

The main Atlantic crossing was carried out by two cables, one for each direction of transmission, with one-way repeaters in flexible housings, designed and built by AT&T and Bell Laboratories. This deep-sea section extended from Oban, Scotland, to Clarenville, Newfoundland, a distance of about 2000 miles.

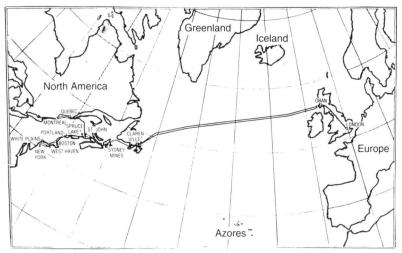

Figure 10.8. Route of first transatlantic telephone cable (1956).

From Clarenville a second cable system continued to Sydney Mines, Nova Scotia, for 340 miles, partly in coastal waters and partly overland, using a single cable with two-way repeaters in rigid housings, of British Post Office Engineering Department design. From Sydney Mines multichannel carrier and microwave radio-relay systems extended the telephone circuits to New York and Montreal.

The completed system provided 35 high-quality telephone circuits from London, 29 to New York and 6 to Montreal. Clearly, a project of this complexity and magnitude involved the contributions of many scientists and engineers at AT&T and Bell Laboratories, the British Post Office, and the U.K. manufacturing industry. Much credit for technical oversight and management of the project must go to the following:

- Dr. Mervin J. Kelly, President, Bell Laboratories
- Sir Gordon Radley, Director General, British Post Office (former Director of Research and Engineer-in-Chief)
- R. J. Halsey, Assistant Engineer-in-Chief (former Director of Research)

The technical problems to be solved were substantial.

On the 2000-mile main transatlantic section, the cable attenuation at the highest working frequency of 164 kHz was 3200 dB (equivalent to 10 multiplied by itself 320 times!), which was compensated by 51 deep-water repeaters, each accommodating 36 telephone circuits. On the Newfoundland–Nova Scotia 340-mile section, the cable attenuation at the highest working frequency of 552 kHz was 1000 decibels, compensated by 16 repeaters, each accommodating 60 two-way telephone circuits. These large gains had to be stabilized, allowing for valve aging and cable loss changes related to temperature variation and other causes, to within a decibel or two, year in and year out. Without Black's invention of negative feedback and the further contributions by Bode and Nyquist, this would have been impossible.

The problem of achieving long life deserves special mention. Clearly all components used in repeaters had to be carefully scrutinized and tested to ensure lives of 20 years or more but the thermionic valves presented unique problems as the main active device. Valve life, and the development of quality assurance techniques, had long been the subject of intensive study in Bell Telephone Laboratories and in the Research Branch of the British Post Office at Dollis Hill, NW London, under the direction of Dr. G. Metson. The latter work, which has been surveyed by M. F. Holmes,[13] had its origins in British Post Office development of North Sea submarine telephone cable systems and later found important applications in transatlantic and transpacific systems.

The problems of laying the cable—and recovering it from ocean depths of a mile or more in the event of a cable or repeater fault—were successfully solved by the British Post Office Research Department's design of a "linear cable engine" mounted in the bow of the BPO telegraph cable ship *H.M.S. Monarch.*[12]

IMPACT OF THE FIRST TRANSATLANTIC TELEPHONE CABLE

Although the number of circuits provided, 35 for commercial use, was small compared with thousands on the later cable and satellite systems, the first transatlantic telephone cable gave a major impetus to traffic growth compared with the shortwave radio telephone systems,

Figure 10.9. British Post Office telegraph cable ship *H.M.S. Monarch*. Laid first TAT cable.

mainly through the greater reliability and better speech quality available on the cable, paving the way for higher-capacity systems in the years that followed.

The first transatlantic telephone cable (TAT1) was followed 3 years later by a telephone cable (TAT2) between Newfoundland and France. In 1961 another milestone was reached in the Atlantic with the laying of a 60-circuit telephone cable from the United Kingdom to Canada (CANTAT 1). This system employed "lightweight" cable in the deep-water sections and "two-wire" rigid repeaters which later, with its derivatives, became a world standard.

The lightweight cable design was originated by Dr. R. A. Brockbank of the British Post Office Research Department; it deserves special mention for its revolutionary impact on submarine cable system design. Earlier cables relied for their strength on heavy external iron-wire armoring; in the Brockbank design the strength element was inside the inner coaxial conductor. This design was cheaper to manufacture, easier to lay because of its lighter weight and greater flexibility, and enabled much longer lengths of cable to be accommodated in the holds of cable-laying ships.

In the 1960s, transistorized repeaters became available, enabling further advances to be made in the performance and capacity of submarine cable systems (Chapter 12). By 1976, 4000-circuit capacity had been achieved on TAT6, and by 1988, 40,000 circuits on an optical fiber transatlantic cable (TAT8) (Chapter 17).

REFERENCES

1. *A History of Engineering and Science in the Bell System: Transmission Technology (1925–1975)*, edited by Gene O'Neill (Bell Laboratories, 1985).
2. O. J. Zobel, "Extensions to the Theory and Design of Electric Wave Filters," *Bell Syst. Tech. J.*, *10*, April 1931.
3. J. B. Johnson, "Thermal Agitation of Electricity in Conductors," *Phys. Rev.*, *32*, July 1928.
4. H. Nyquist, "Thermal Agitation of Electric Charge in Conductors," *Phys. Rev.*, *32*, July 1928.
5. Ref. 1, Chapter 4.
6. H. Nyquist, "Regeneration Theory," *Bell Syst. Tech. J.*, January 1932.
7. H. W. Bode, "Relations Between Attenuation and Phase in Feedback Amplifier Design," *Bell Syst. Tech. J.*, January 1932.
8. O. E. Buckley, "Evolution of the Crystal Wave Filter," *J. Appl. Phys.*, *8*, January 1937.
9. Post Office Electrical Engineers Journal, 1906–1981, 75th Anniversary, *74*, Pt. 3, October 1981.
10. R. A. Brockbank and C. A. Wass, "Non-Linear Distortion in Transmission Systems," *J. IEE*, *92*, 1945.
11. A. H. Mumford, "The London–Birmingham Coaxial Cable System," *Post Off. Electr. Eng. J.*, *30*, Pt. 3, October 1937.
12. Post Office Electrical Engineers Journal: The Trans-Atlantic Telephone Cable Number, *49*, Pt. 4, January 1957.
13. M. F. Holmes, "Active Element in Submerged Repeaters: The First Quarter Century," *Proc. IEE*, *123*, No. 10R, October 1976.

11

The First Microwave Radio-Relay Engineers

WHAT ARE MICROWAVES?

Microwaves are electromagnetic radio waves with frequencies from about 1000 MHz (30-cm wavelength) to 30,000 MHz or 30 GHz (1-cm wavelength); radio waves of even shorter wavelengths, from 30 to 3000 GHz (10-mm to 0.1-mm wavelength), are referred to as "millimetric" waves. Beyond the millimetric waves are the infrared and visible-light regions of the electromagnetic wave spectrum.

Microwaves are remarkably suitable for the purposes of point-to-point telecommunication because they can be sharply beamed by directional aerials of mechanically convenient dimensions, a property that also made them very effective for the radar systems developed during World War II. Since microwaves do not readily diffract around the curved surface of the Earth, they are mainly useful over "line-of-sight" paths between transmitter and receiver. An exception to this generalization is possible by the mechanism of "tropospheric scattering" in which microwaves can be propagated to some extent beyond the visible horizon by wave scattering from regions of the lower atmosphere or troposphere characterized by nonuniform humidity and temperature.

The discovery of the existence and useful properties of microwaves owes much to Heinrich Hertz and his experiments in 1886 which confirmed James Clerk Maxwell's predictions of the identity of radio waves and light. Although Hertz's experiments with spark generators and gaps, resonant loops, and parabolic reflectors were probably conducted on frequencies up to some 600 MHz (50-cm wavelength), they nevertheless laid the foundation for developments that were to prove of great value. As already noted (Chapter 2), Hertz did not envisage the use of microwaves for long-distance communication because they "would require a mirror as large as a continent!" Hertz's work has been recognized by the use internationally of the term "Hertzian waves," and by the French "Faisceau Hertzien" to describe a beamed microwave link. By the 1920s other experimenters using similar techniques had generated frequencies up to 15 GHz and beyond, but practical application of microwaves had to await other events.

PREWAR VHF/UHF AND MICROWAVE RADIO-RELAY DEVELOPMENTS

In the 1930s the British Post Office made extensive use of VHF radio links to provide single telephone circuits, or groups of 6, 12, or 24 frequency-division multiplex circuits between the mainland and offshore islands, e.g., off the coasts of Scotland and Wales, and to the Channel Islands. These links were engineered by the Radio Experimental Branch of the Post Office at Dollis Hill, London, under the direction of A. H. Mumford, H. T. Mitchell, and D. A. Thorn.

Frequencies in the range 50 to 100 MHz were generally used, the lack of commercially available valves precluding the use of higher frequencies at the time. The radio equipment initially used amplitude modulation of the carrier wave; a later development using frequency modulation is referred to below. The aerials were commonly arrays of half-wavelength dipoles, or a scaled-down version of the convenient and economical Bruce rhombic aerial used in shortwave overseas services. A 65-MHz link providing 15 telephone circuits over a 40-mile path between Scotland and Northern Ireland was brought into service by the British

Figure 11.1. British Post Office VHF/UHF offshore multichannel telephony links (1938).

Figure 11.2. The first experimental microwave link (STC/LMT) across the English Channel (1931).

Post Office in 1936. The link to the Channel Islands involved an 85-mile oversea path between Chaldon, Dorset, and Guernsey, Channel Islands; frequencies of 37.5 and 60 MHz were used for the commercial telephony service, enabling valuable radio-wave propagation data to be obtained by comparing transmission over two widely different radio frequencies.[1]

This four-channel telephony link was closed down during the war years; its reinstatement after the end of the war played a major role in restoring the morale of the Channel Islanders.

Similar developments using the VHF/UHF spectrum occurred in the United States, notably by the Bell Laboratories. An early example was a single-channel two-way telephony link established between Green Harbor and Provincetown, Massachusetts, in 1934, using frequencies of 63 and 65 MHz. An amplitude-modulated FDM link, with five equipped telephone channels, and using frequencies of 156 and 161 MHz was set up on a 26-mile path between Cape Charles, Maryland, and Norfolk, Virginia, in 1941.[2]

These early essays in FDM multichannel telephony operation of VHF/UHF radio links provided valuable experience and a degree of useful commercial service. However, they revealed the limitations of amplitude modulation arising from inadequate linearity of the modulation and amplification processes, which gave rise to interchannel crosstalk. It also became clear that the frequency space available in the VHF/ UHF spectrum was insufficient to allow for traffic growth and any large-scale exploitation. These considerations prompted research to find more effective modulation methods and means for using the spectrum above 1000 MHz—the microwave region.

FREQUENCY MODULATION PROVIDES A SOLUTION

Major E. Armstrong's invention of frequency modulation had proved of great value in VHF broadcasting, notably by improving the signal-to-noise ratio and avoiding nonlinear distortion (Chapter 8). The first application of FM to FDM radio telephony systems arose in the United Kingdom in the late 1930s as a means of combatting excess noise on a 12-channel AM VHF link on the 65-mile path between Holyhead, Anglesey, and Douglas, Isle of Man. This link suffered at times from interference caused by salt-encrusted insulators on a high-voltage power line near the receiving aerials. Post Office Radio Experimental Branch engineers J. H. H. Merriman and R. W. White at the BPO Castleton Laboratories, South Wales, designed and built FM equipment suitable for a 12-channel FDM 60–108 kHz group.[3] The experiment proved a dramatic success; not only was the power-line interference greatly reduced without increasing the transmitter power, but interchannel crosstalk was virtually eliminated. (The author made a small contribution to the latter by a theoretical study enabling the phase linearity requirements for low-distortion transmission of a frequency-modulated signal to be approximately determined; a similar but more precise study was later made by L. L. Lewin and colleagues of the Standard Telecommunication Laboratories, U.K.[4]).

This successful use of FM for FDM telephony radio-relay systems had an impact far beyond the specific application in which it was first tried: it became the norm for microwave radio-relay systems bearing 1000 or more telephone channels and for television relaying, standardized by the International Radio Consultative Committee and used throughout the world.[5,17]

A PIONEERING MICROWAVE RADIO-RELAY
EXPERIMENT (1931)

To the engineers of the Standard Telephones and Cables Ltd. in England and Les Laboratoires de Matériel Téléphonique in Paris must go credit for the first pioneering experiment in the use of microwaves, heralding the shape of things to come. This experiment in March of 1931 involved a 1720-MHz link over a 21-mile path across the English Channel between Calais and St. Margaret's Bay, near Dover, providing a single two-way telephone channel.[6]

This link used front-fed parabolic reflector aerials some 3 meters in diameter, with a power gain of 26 decibels; this large gain enabled the link to function with a transmitter power of less than 0.5 watts. The transmitter consisted of an oscillator invented by Barkhausen-Kurz in 1919—a triode valve in which the grid was held at a high positive potential and the anode at a low one, the electrons oscillating between grid and anode at an extremely high frequency, largely independent of the external circuit. It is not clear whether the modulation was in amplitude or frequency, perhaps a combination of both.

The experiment was notable not only for its pioneering use of microwaves, but also because it revealed propagation variations caused by the rise and fall of the tide in the English Channel and its effect on the phase of the sea-reflected wave—a phenomenon that had to be allowed for in the design of the first commercial Cross-Channel microwave link in the 1950s.

THE IMPACT OF THE WAR YEARS
ON MICROWAVE DEVELOPMENT

The wartime development of radar—the fascinating story of which is outside the scope of this book—gave an impetus to the evolution of microwave techniques and devices in many areas including aerials and waveguides, klystron oscillators, pulsed magnetrons, silicon crystal mixers, microwave filters and other components, and intermediate-frequency (typically 45 MHz) valve amplifiers.

The klystron was one of the most valuable contributions to microwave radio-relay system development. It was invented in 1939 by the brothers R. H. and S. F. Varian of Stanford University, and used the principle of velocity modulation of an electron stream by a varying applied voltage generated across a gap in a microwave cavity resonator. Velocity variation caused electrons to "bunch," faster electrons overtaking slower ones, as they progressed along the stream. Energy could be abstracted from the bunches by a second suitably placed microwave cavity resonator, thus creating a microwave amplifier. Similar principles could be applied in a single cavity "reflex" system, creating a microwave oscillator.[7]

The war also stimulated the development of time-division multiplex telephony techniques using width or position modulation of trains of pulsed microwave carriers generated by klystrons or magnetrons, foreshadowed in their 1938 patent by E. M. Deloraine of LMT in France and A. H. Reeves of STC in England.[8] In England a military 8-channel system designed in 1942 used a 5-GHz magnetron, and in Germany a 10-channel system designed by Lorenz used initially a 1.3-GHz magnetron and later a 1.3-GHz klystron.[5]

As the wartime development of radar moved up the spectrum from 3 GHz (10 cm—"S" band) to 10 GHz (3 cm—"X" band), the need for low-loss transmission of microwave power stimulated the evolution of waveguides—generally in the form of rectangular copper tubes with dimensions related to the wavelength. Coaxial cables tended to be too lossy at such frequencies, and open-wire lines radiated too freely. At microwave frequencies, conventional circuit elements such as inductors and capacitors became vanishingly small, and here waveguide components became available to take their place in filters and power dividers/combiners.

Lord Rayleigh had shown in 1897 that certain solutions of Maxwell's equations (Chapter 2) predicted that the transmission of electromagnetic waves through hollow conducting tubes was feasible, and that a variety of modes involving various combinations of magnetic and electric field patterns was possible. Pioneering work in Bell Laboratories in the 1930s, notably by G. C. Southworth, J. R. Carson, S. P. Mead, and S. A. Schelkunoff, established the basic criteria for waveguide design, including the cutoff frequencies and losses related to imperfect conductivity. At Massachusetts Institute of Technology, W. L. Barrow was also conducting a program of research on waveguides, including the use of resonant waveguide cavities as filter.

Thus, by the end of the war in 1945, the groundwork had been prepared for the development of the high-capacity microwave radio-relay systems that were to become of major importance in the telecom-munication networks of the United States, Europe, Japan, and other countries.

POSTWAR MICROWAVE RADIO-RELAY SYSTEM RESEARCH IN THE BELL LABORATORIES

System studies by AT&T in the later years of the war had indicated a good possibility that microwave radio-relay systems could provide an economic alternative to coaxial cable systems for both multichannel telephony and television, with the advantage of speedier provision in some cases. A substantial program of research and system development was initiated in Bell Laboratories on the basis of this forward-looking study.[2]

Central to the research program were studies of microwave propa-gation over typical land and sea line-of-sight paths, with the objectives of determining the optimum region of the spectrum for radio-relay system use and the rules for system design, recognizing that microwaves were subject to a degree of fading under certain weather conditions and that attenuation caused by heavy rainfall might affect the higher microwave frequencies. These studies indicated the suitability of fre-quencies around 4 GHz (7.5-cm wavelength) for the first systems, with the possibility of later using other frequencies in the range 2 to 10 GHz. An important consideration in system planning was the necessary clearance of the direct ray path over intervening terrain to avoid exces-sive fading caused by the ground or sea reflected rays and ray-bending from abnormal atmospheric refraction—important because it governed the height and cost of towers needed to support the microwave aerials. The propagation studies of microwave fading had to be carried out over periods of a year or more to provide statistical data enabling micro-wave repeater design to incorporate sufficient reserve gain and signal-to-noise margin to maintain the necessary overall standards of perfor-mance of long radio-relay systems.

Harald T. Friis (b. 1893)

Concurrently with the work on microwave propagation, the Bell Laboratories in Holmdel, New Jersey, were pursuing fundamental studies of microwave components, aerials, repeater, and system design under the leadership of H. T. Friis, about whom it was said:

> Friis, an outstanding researcher and radio engineer, in addition to innumerable personal contributions, served as mentor to a generation of Bell Laboratories radio scientists and engineers. He was as much loved and respected for his human qualities as for his technical acumen. To those who knew him, the enormous microwave-radio network is a fitting memorial to his work.[2]

Friis has written a fascinating autobiography, *Seventy-Five Years in an Exciting World*.[9] He was born in a small town in Denmark in 1893, one of a family of ten children, four of whom died in infancy. His father was a brewer who died at the age of 48, leaving his mother with six children and a heavily mortgaged brewery to run. Family life was conducted in very straightened circumstances, and Harald was apprenticed to a blacksmith for a time. During his Technical College studies he was fortunate to study under Professor Niels Neilson, the famous mathematician; he also worked under Valdemar Poulsen, the inventor of the Poulsen Arc and a magnetic tape system for recording speech.

Friis joined the Bell Laboratories in 1919; one of his first tasks was to work on a "double detection" radio receiver, i.e., one in which an incoming modulated radio-frequency carrier is first translated to an intermediate frequency for convenient amplification and before demodulation—a principle that he describes as "one of the most important contributions to the radio art," attributed to Lucien Levy in France. His early work was on long-wave reception, moving on later to shortwave radio, including single-sideband and MUSA systems (Chapter 7).

Friis had much to do with the setting up of the Holmdel field laboratory and the creation of a down-to-earth, informal environment in wooden huts. The author visited Holmdel shortly after the end of the war, at a time when the British Post Office was beginning to plan new laboratories, and asked Friis, "How big do you think an ideal research laboratory should be?" Harald thought for a while and said, "Perhaps about 100 people, I could then handpick them for the jobs to be done." Today's laboratories with staff measured in thousands, housed in air-conditioned glass-fronted multistory buildings, are a long way from Friis's ideal! Not even Holmdel was immune to visits from high authority; on one occasion the author happened to be in the small conference room in one of the wooden huts and noticed that, on a blackboard full of scrawled circuit diagrams, mathematical symbols, and triple-integrals, a firm clear hand had written "dollars per message channel per mile"—on inquiry he was told that Dr. Kelly, then President of the Laboratories, had just paid a visit, and evidently determined to make clear to his scientist/engineers what their real objective was!

In the field of microwave radio, Friis will be remembered in particular for two important contributions. The first stated a law, as important for microwave system design as Ohm's law is for electric circuits:

$$P_r/P_t = (A_r \times A_r)/(\lambda^2 \times d^2),$$

where P_r and P_t are the received and transmitted powers, A_t and A_r are the effective areas of the transmitting and receiving aerials, λ is the wavelength, and d is the distance between the transmitting and receiving aerials. All good microwave engineers should have this equation firmly in their heads if not engraved on their hearts!

The second notable contribution was the horn-reflector aerial invented by Friis and A. C. Beck at Holmdel, and based on earlier work by G. C. Southworth on the radiating properties of conical horns.

The horn-reflector aerial had a number of valuable properties: it was effective over a wide frequency range, typically from 2 to 10 GHz, in a single aerial, with good impedance matching over the whole of this range, enabling it to be used simultaneously for several radio-

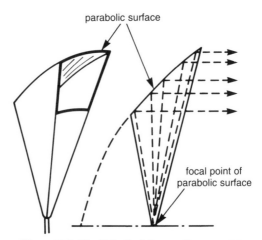

Figure 11.3. The Friis–Beck horn-reflector aerial.

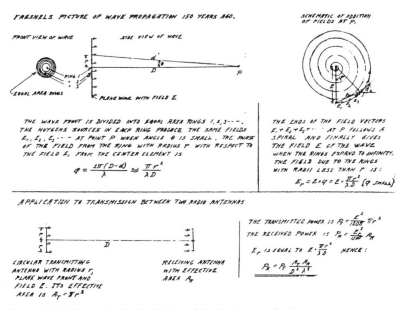

Figure 11.4. Derivation by H. T. Friis of the fundamental microwave transmission formula: $P_r/P_t = (A_r \times A_r) / (\lambda^2 \times d^2)$

frequency channels. It could accommodate both vertical and horizontal polarization, and it offered good protection against unwanted signals from the side and behind, an important feature in multiple-hop radio-relay system design.

So successful was the horn-reflector aerial that it was used widely for radio-relay systems throughout the United States, Europe, and other parts of the world. A steerable version at Holmdel was used for wave propagation research and became a prototype for the much larger radome-protected steerable horn-reflector aerial at Andover, Maine, for the TELSTAR communication satellite experiment in 1961 (Chapter 15).

Friis's eminent colleague J. R. Pierce has written:

> I have known cleverer inventors, more abstruse scientists, deeper mathematicians, better politicians, and executives of higher degree. I have known no other man who has left as deep and profitable impress on those who have worked for or with him, or who has had a clearer insight or a surer success in the work he has undertaken.[9]

RADIO-RELAY DEVELOPMENT
IN THE AT&T/BELL SYSTEM

From about 1945, microwave radio-relay system development in the United States received a strong impetus from the urgent need of AT&T to provide television relaying facilities throughout the country to meet the needs of the powerful broadcasting companies—the Columbia Broadcasting System, the National Broadcasting System, and the Radio Corporation of America. The economics of television program provision, and the needs of the advertisers on whom the whole financial structure depended, alike required the ability of each broadcasting company to link its many television stations so that a single program or advertisement could be transmitted throughout the nation. Any failure by AT&T to provide nationwide television relaying facilities in good time would have seriously jeopardized its virtual monopoly over long-distance interstate communication by opening the door to competitors.

AT&T already had a nationwide network of coaxial cables, each transmitting 1000 or more telephone circuits. However, the coaxial cable network had been designed for maximum efficiency in the transmission of multichannel telephony and was not technically well suited to television relaying without considerable modification.

Fortunately—or by good foresight—the microwave research at Holmdel had reached a point where a multihop radio-relay field trial could be planned, built, and tested. The route chosen was from New York to Boston, a distance of 220 miles, in five hops averaging 27.5 miles. The frequencies used for two microwave carriers in each direction of transmission were in the 3700–4200 MHz band eventually assigned by the Federal Communications Commission for "common-carrier" use.

Frequency modulation of the microwave carriers was used, both for television and for FDM multichannel telephony, the primary emphasis of the experiment being on television.

The aerials on this experimental system—designated "TDX"—were of the "delay-lens" type designed by Dr. Koch of the Bell Laboratories. Waveguide branching units, invented by W. Tyrell of the Laboratories, enabled a number of microwave carriers on different frequencies to be combined for transmission via a common aerial, or separated for reception.

Repeater design was based on conversion of the incoming microwave carrier, via a crystal mixer and klystron local oscillator, to an intermediate frequency of 65 MHz at which amplification and gain regulation to allow for fading took place. The intermediate-frequency signal was then converted in a crystal mixer to an outgoing microwave carrier displaced 40 MHz from the incoming carrier frequency. This frequency shift was necessary to avoid undesirable feedback from the output to the input of the repeater via the limited back-end directivity of the transmitting and receiving aerials. The outgoing microwave carrier was raised in a velocity-modulation klystron amplifier to a power of about 1 watt.

Overall performance tests in 1947 demonstrated that television picture quality was virtually unimpaired after transmission over the TDX system, and some 240 FDM telephone circuits of satisfactory quality were possible.[11]

The basic design concepts of the TDX experimental microwave radio-relay system were demonstrably sound and were followed in the TD2 operational system that was eventually used by AT&T throughout the United States for television and multichannel telephony. Initial changes in design incorporated in TD2 included the use of the Friis horn-reflector aerial in place of the delay-lens aerial, and a 4-GHz triode valve power amplifier in place of the klystron, with the advantages of lower operating voltages and longer life.[12]

The 4-GHz triode, designed Dr. J. B. Morton of the Laboratories, took the upper limit of operation of the triode valve to a point much higher than had earlier been achieved. It was a remarkable tour de force, requiring extremely fine mechanical tolerances on the spacings between the cathode, grid, and anode, and precise control of the manufacturing processes.

Over the period 1950 to 1980, a series of design improvements were made in the TD2 4-GHz radio-relay system, and eventually 1800 FDM telephone circuits were achieved on each radio carrier, and up to 11 carriers on a route. Microwave systems on frequencies of 6 GHz (TH) and 11 GHz (TL) were also developed.

These systems used frequency modulation, but more recent systems have employed single-sideband amplitude modulation for FDM telephony, with significant gains in efficiency of spectrum utilization.

The demonstrated success of the TDX experimental microwave radio-relay system and its operational successors stimulated similar developments in the United Kingdom, Europe, and Japan, with varia-

tions of design according to the various national requirements and resources.

MICROWAVE RADIO-RELAY SYSTEM DEVELOPMENT
IN THE UNITED KINGDOM

When the study of radio-relay by the Radio Experimental Branch
of the British Post Office Engineering Department was resumed shortly
after the end of the war, there was a lack of commercially available
microwave amplifiers in the United Kingdom, and attention was first
directed to the use of frequencies around 200 MHz for which triode
valve amplifiers with power outputs of a few watts could be obtained.

An experimental radio-relay system operating at 200 MHz was
designed and built by the Radio Branch of the Engineering Department
(ED) between the BPO Research Station at Dollis Hill, NW London, and
ED Laboratories at Castleton, S. Wales. This served primarily as a test-
bed for radio-relay system experimental work but was spurred on by
the need for a temporary television link between London and Wales
for the opening of the British Broadcasting Company television transmitter at Wenvoe, near Cardiff, in 1952.

This five-hop radio-relay system was unique in that it was designed
to use a single frequency throughout, with Bruce-type horizontal rhombic aerials in pairs on opposite sides of hilltop sites, so located that
diffraction loss over the hilltop avoided interaction between the outgoing and incoming 200-MHz signals.

The London–Wenvoe system was also of interest in that it marked
one of the earliest uses of frequency modulation for the radio-relaying
of television signals in the United Kingdom.[13] It demonstrated clearly
the advantages of FM over AM for such applications, and this became
the accepted type of modulation for future radio-relay systems.

Figure 11.5. The invention of the traveling-wave tube microwave amplifier. (Top)
John Pierce (left) and Rudolf Kompfner. (Bottom) From Rudolf Kompfner's notebook.
The notebook entry dated Nov. 12, 1942, in which an output resonator is replaced
by a section of helix, leads to the astonishing conclusion: "A completely untuned
amplifier? Would it work? Are the electrons in the output region not moving parallel to
the equipotential surfaces of the line? If so, then there can be no amplified shot noise."

12. 11. 42

signal

A completely contained amplifier!

output

Would it work? Are the electrons in the output region not moving
parallel to the equipotential surfaces of the line? If so, then
there can be no amplified shotnoise

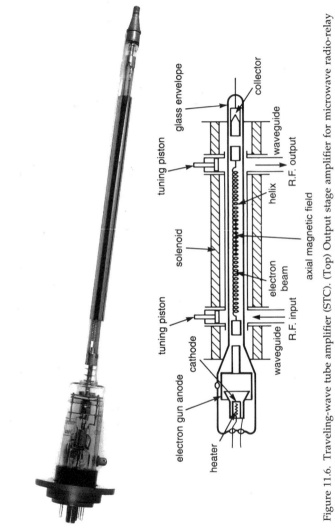

Figure 11.6. Traveling-wave tube amplifier (STC). (Top) Output stage amplifier for microwave radio-relay system. (Bottom) Traveling-wave tube amplifier sectional diagram.

A 900-MHz radio-relay system for television, built under Post Office contract by the General Electric Company (U.K.) to link the television switching center at the PO Museum Exchange in London with the BBC television transmitter at Sutton Coldfield in Birmingham, opened in 1949.[14]

By the 1950s, and with an eye on developments in the United States, it had become clear that the future development of radio-relaying, both for multichannel telephony and for television, lay in the microwave region of the radio spectrum where sufficient frequency space was available for large-scale exploitation. The first commercial development of microwave radio-relay systems in the United Kingdom was carried by Standard Telephones and Cables Ltd. (STC) with their 4-GHz system, using traveling-wave tubes as amplifiers, and by the General Electric Co. (U.K.), with a 2-GHz system using disk-seal triode valve amplifiers.

The STC 4-GHz radio-relay system, built under contract to the British Post Office for the Manchester–Kirk o'Shotts, Scotland, television link and opened for service in 1952, was the first microwave radio-relay system in the world to use traveling-wave tubes on an operational basis.[15] The British Post Office, recognizing the design and manufacturing problems of triode valves with close-spaced electrodes, had earlier arranged sponsorship of TWT development with STC via the U.K. Co-ordination of Valve Development Committee. This development work, directed by D. C. Rogers of STC at their Ilminster, Somerset Laboratories, resulted by 1949 in the highly successful 1-watt-output 4-GHz traveling-wave tube amplifier first used in the Manchester–Kirk o'Shotts radio-relay system.

The traveling-wave tube proved of major importance in the development of microwave radio-relay systems: it offered wide signal bandwidth and could be readily designed for operation at frequencies of up to at least 15 GHz without involving unduly critical mechanical tolerances. Designs for low-input noise amplifiers, and also for high-output power amplifiers became possible. Important applications in communication satellite systems and in military radio surveillance systems also followed. The inventor of the traveling-wave tube was Rudolf Kompfner, when working at the Clarendon Laboratory, Oxford, during the war years—and it was through our joint membership in the U.K. Co-ordination of Valve Development Committee that the author became aware of the possibilities of the traveling-wave tube for radio-relay systems.

THE INVENTION OF THE TRAVELING-WAVE TUBE

Rudolf Kompfner (1909–1977)

Rudolf Kompfner was born in Vienna, where he trained as an architect at the Technische Hochscule. He came to England in 1934 and worked as an architect for a time, but was interned as an alien at the outbreak of war in 1939. On his release he joined an Admiralty group working on tube research for radar at the University of Birmingham, under the direction of Professor L. M. Oliphant. It was in this group that Randall and Boot invented the multicavity pulsed magnetron that played such a vital role in the radar systems that helped the Allies win the war.

Kompfner's invention of the traveling-wave tube is told, with characteristic modesty, in his book of the same title.[16] His work with the Birmingham group had initially been concerned with attempts to reduce the inherent noise in klystron amplifiers, which necessarily involved a study of the principles of velocity modulation of electron beams, first published by O. and A. Heil in 1935. The aim was a low-noise amplifier that would improve the sensitivity of radar receivers above that achievable by crystal detector mixers.

However, the traveling-wave amplifier came about not directly from the klystron work, but rather from an attempt to devise a high-frequency oscilloscope by slowing down the signal wave so that it could interact with, and deflect, an electron beam. The slow-wave structure Kompfner chose was a wire helix—a stroke of genius because it was by no means apparent that a microwave would follow the turns of the helix. By November, 1942, the possibility of using such a structure as an amplifier, but with a hollow electron beam surrounding the wire helix, had occurred to him and his notebook records: "A completely untuned amplifier?" (Figure 11.5). After many false starts, and an eventual realization that the microwave field on the helix had an axial component that could interact with an axial electron beam and produce velocity-modulation "bunching," success was achieved in November, 1943, with a tube that produced a modest power gain of 1.5 times. By 1944 a power gain of 10 times had been achieved in a traveling-wave tube with less than half the noise level of typical crystal mixers—the principle of the traveling-wave tube had been fully and clearly demonstrated.

In 1944 Kompfner moved on to the Clarendon Laboratory, Oxford, where he continued work on the traveling-wave tube, aiming at a theory that would enable design to be optimized and performance predicted.

A visit to the Clarendon Laboratory in that year by Dr. J. R. Pierce of the Bell Laboratories produced a partnership in the study of the traveling-wave tube that yielded a number of valuable results, notably Pierce's more precise theory that enabled gain and other performance characteristics to be accurately predicted. This theory revealed that loss in the helix was actually beneficial in that it enabled higher gain to be achieved with stability; it also indicated that the useful bandwidth could exceed an octave.

The Bell Laboratories did not lose sight of the advantages of the traveling-wave tube: after the TD2 4-GHz triode-based radio-relay system, the next higher frequency systems used traveling-wave tubes.

Kompfner joined the Bell Laboratories at Holmdel in 1951, where his unique powers of scientific leadership were of great value to teams working on communication satellites, lasers, and radioastronomy. In 1973 he began a new career as a professor of engineering, sharing his time between Oxford and Stanford universities. He received many awards and distinctions, including honorary doctorates from the universities of Vienna and Oxford, and in 1976 the American President's Award for Achievement in Science—a remarkable record for someone who had no formal training in science and indeed did not embark on a scientific career until he was well into his thirties.

It was said of Rudolf Kompfner that "he was a man of perfect integrity, great charm and boundless generosity"; the author is proud to have known him as a friend and a wartime colleague.

THE INTERNATIONAL STANDARDIZATION OF RADIO-RELAY SYSTEMS

Beginning in the early 1950s, and following the lead of the United States, the use of microwave radio-relay systems by the Post, Telephone and Telegraph Administrations throughout the world began to expand on a large scale—with distinctive contributions to the engineering of such systems by the United Kingdom, France, Germany, and Japan.[5]

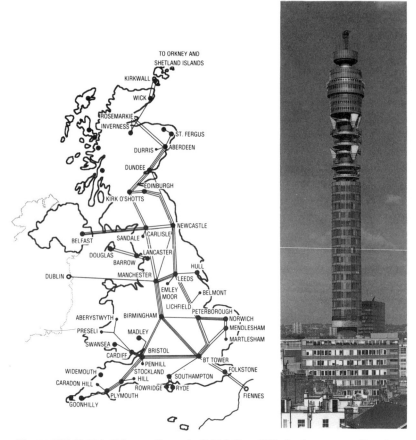

Figure 11.7. British Telecom (formerly British Post Office) microwave radio-relay trunk network (1980). (Left) Microwave radio-relay trunk network in the United Kingdom. (Right) British Telecom Tower (London)—focal point of the network.

For radio-relay systems to provide satisfactory transmission of telephony, television, and data over long distances, and to facilitate their interconnection with one another and with line systems, certain common characteristics are necessary. In particular, the radio-frequency channels in the spectrum, and the modulation characteristics of the radio carrier waves must be defined.

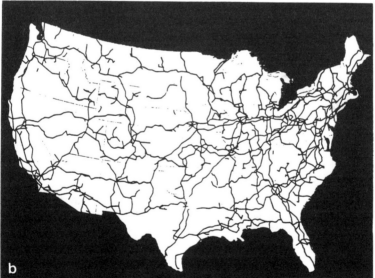

Figure 11.8. Growth of microwave radio-relay networks in Europe and the United States. (Top) Microwave radio-relay (TV) network in Europe (1961). (Bottom) Microwave radio-relay (TD2) network in the United States (1980).

International agreement on preferred standards has been achieved through the work of the International Radio Consultative Committee of the International Telecommunication Union in Geneva, through its Study Group IX (Radio-Relay systems).[17] These standards have not only greatly facilitated the setting up of radio-relay links across national frontiers, they have also given a valuable lead to equipment manufacturers enabling them more efficiently to meet the needs of a growing world market. The standards are a remarkable example of the ability of engineers from a wide range of countries—including the USSR—to reach agreement on common solutions to technical problems, when these are based on scientific and engineering facts and agreed operational needs.

It is satisfying to record that one of the first applications of the CCIR radio-relay system standards was to the 4-GHz multichannel telephony and television link across the English Channel established by the British Post Office and the French PTT Administration in 1959, using equipment made by STC in England and LMT in France—an interesting outcome from that pioneering experiment across the same path in 1931!

REFERENCES

1. Chairman's Address by A. J. Gill, Wireless Section of Institution of Electrical Engineers, *J. IEE*, *84*, No. 506, February 1939.
2. *A History of Engineering and Science in the Bell System: Transmission Technology (1925–1975)*, edited by Gene O'Neill (Bell Laboratories, 1985), Chapter 7.
3. J. H. H. Merriman and R. W. White, "Frequency Modulation Tests on the Holyhead–Douglas Ultra-Short Wave Radio Link," Post Office Engineering Department Radio Report No. 806, September 1942.
4. L. Lewin *et al.*, "Phase Distortion in Feeders," *Wireless Engineer*, *27*, 1950.
5. Helmut Carl, *Radio-Relay Systems* (Macdonald and Co., London, 1966).
6. A. G. Clavier and L. C. Gallant, "Anglo-French Micro-ray Link between Lympne and St. Inglevert," *Electr. Comm.*, *12*, 1934.
7. R. H. and S. F. Varian, "A High-Frequency Oscillator and Amplifier," *J. Appl. Phys.*, *10*, 1939.
8. E. M. Deloraine, "Evolution de la Technique des Impulsions Applique au Telecommunications," *Onde Electr.*, *33*, 1954.
9. H. T. Friis, *Seventy-Five Years in an Exciting World* (San Francisco Press, 1971).
10. H. T. Friis and A. C. Beck, Horn-Reflector Aerial, U.S. Patent No. 2,236,728, filed October 1936, issued March 1941.

11. G. N. Thayer *et al.*, ''A Broad-band Microwave Relay System between New York and Boston,'' *Proc. IRE, 37*, 1949.
12. A. C. Dickieson, ''The TD2 Story—from Research to Field Trial,'' *Bell Lab. Rec., 45*, October 1967.
13. W. J. Bray, ''Post Office Contributions to the Early History of the Development of Television in the United Kingdom,'' *British Telecom J., 5*, No. 1, January 1987.
14. R. I. Clayton *et al.*, ''The London–Birmingham Television Radio-Relay Link,'' *Proc. IEE, 98*, Pt. 1, 1951.
15. G. Dawson and L. L. Hall, ''The Manchester–Kirk o'Shotts Television Radio-Relay System,'' *Proc. IEE, 101*, Pt. 1, 1954.
16. Rudolf Kompfner, *The Invention of the Traveling-Wave Tube* (San Francisco Press, 1964) and *Wireless World* (UK), *52*, November 1946.
17. W. J. Bray, ''The Standardization of International Microwave Radio-Relay Systems,'' *Proc. IEE, 108*, Pt. B, No. 38, 1961.

12

The Inventors of the Transistor and the Microchip

A Worldwide Revolution in Electronics

SIGNIFICANCE OF THE TRANSISTOR AND THE MICROCHIP

The inventions of the transistor in the 1940s and the planar integrated circuit—the microchip—in the 1960s began a development in electronics that was to have a wide-ranging, profound, and continuing impact on telecommunications, sound and television broadcasting, and computing throughout the world. They enabled electronic equipment to be made more compact, more reliable, and lower in cost and power consumption than was possible using thermionic valves. The microchip in particular enabled circuit operations of far greater complexity to be performed reliably, rapidly, and economically, greatly enhancing the capability of computers to calculate, the service functions available in electronic telephone exchanges, and the quality of color television broadcasting. The transistor and the microchip facilitated the design of larger-capacity land and submarine cable systems and the design of communication

177

satellites. They made possible a vast new range of customer equipment for computing, communicating, and broadcasting; they gave rise to new electronic industries and changed old ones beyond recognition.[1]

Transistors exploit the property of monocrystals, in particular of one or other of the elements germanium and silicon (both in Group 4 of the periodic table), to conduct electricity by way of two separate carriers whose presence depends on the extent to which crystals have been made impure, at very low levels, by the addition of atoms of Group 3 (notably boron or indium) or Group 5 (notably phosphorus and arsenic).

By suitable design, active circuit elements such as rectifiers, amplifiers, and switches can be created in small physical sizes, limited only by the amount of electrical power to be handled, and with long life expectation. Thermionic valves involving the emission of electrons from heated filaments or cathodes are necessarily bulky, and when mass-produced, tend to be of limited life.

Initially transistors tended to be used individually on "wired-circuit" boards, connected to other circuit elements such as resistors and capacitors by handmade or preformed wiring involving operations by human beings—a process that set a limit to miniaturization and also reliability.

The next, and giant, leap forward was to integrate all of the components on a single slice of material, usually silicon, giving rise to the planar integrated circuit or microchip. This brilliant concept paved the way for almost unlimited miniaturization—with many thousands, even millions, of transistors or other circuit elements on a few square millimeters of silicon.

The microchip made possible highly automated manufacturing processes, offering greatly enhanced reliability and cost reductions achieved by large-volume production.

THE PREHISTORY OF THE INVENTION OF THE TRANSISTOR

Solid-state devices in the form of crystals, e.g., of galena and carborundum, had been used in receivers for detection of modulated carrier waves from the beginning of the radio art, as had copper-oxide rectifiers for converting alternating current to direct current. There were reports in the British and American technical press (*Wireless World*, May, 1920, and *QST*, March, 1920), describing a radio receiver designed by Dr. E. W. Pickard in which an oscillating crystal diode in a circuit containing a DC power source was used for heterodyne reception of continuous-wave carrier signals. The discovery of crystal diode oscillation has been attributed to Dr. W. H. Eccles, some 10 years earlier.[3] And "crystals" had been used as mixers at the front end of radar receivers and microwave repeaters. But these were all essentially diodes. What was needed was a solid-state device capable of amplification.

It has been said that "had it not been for De Forest's Audion triode valve, the transistor would have been invented much earlier than 1948." Be that as it may, there were signs from the 1920s onwards that the triode valve might not be the only possible device for generating and amplifying high-frequency oscillations.

In 1923 Dr. Julius E. Lillienfeld (pioneer of the electrolytic capacitor) in Canada demonstrated a "tubeless" radio receiver in which solid-state devices acted as oscillators and amplifiers. Between 1925 and 1928 Lillienfeld applied for Canadian patents for devices that had the geometry of today's insulated-gate field-effect transistors. The German engineer O. Heil (inventor of the Heil velocity-modulation tube) filed a British patent in 1934 for a device based on a copper-oxide rectifier with an added electrode. In 1936 Holst and van Geel of the Philips Laboratories in Holland patented a type of bipolar device; van Geel continued the work on devices during the war years, filing other patents in 1943 and 1945.[4]

However, none of these efforts came to fruition, perhaps because of the lack of in-depth scientific knowledge of the phenomena involved, of detailed knowledge of the materials technology and of the engineering skills needed to evolve practical designs. It remained for scientists and engineers of the Bell Telephone Laboratories in the United States to fill this gap.

THE INVENTION OF THE TRANSISTOR: THREE MEN WHO CHANGED OUR WORLD

This was the proud title that appeared on the front page of the *Bell Laboratories Record* in December, 1972,[5] a commemorative issue describing events that led up to the invention of the transistor in 1948 and the developments that followed in the next 25 years. The three men, then research physicists in the Bell Laboratories and who jointly received the Nobel Prize in Physics in 1956, were Dr. John Bardeen, Dr. Walter H. Brattain, and Dr. William Shockley.

John Bardeen, who joined Bell Labs in 1945, was coinventor with Walter Brattain of the point-contact transistor. He received his early scientific education at the universities of Wisconsin and Princeton, later becoming President of the American Physical Society and a member of the U.S. President's Science Advisory Committee. He received many honors, including a second Nobel Prize in 1972 for work on superconductivity, and became Professor of Electrical Engineering and Physics at the University of Illinois.

Walter Brattain took M.A. and Ph.D. degrees at the universities of Oregon and Minnesota. He joined Bell Labs in 1929, working on the surface properties of solids and semiconductor materials. His studies on the latter led him, jointly with John Bardeen, to the invention of the point-contact transistor. His many honorary awards included D.Sc. degrees from the universities of Portland and Minnesota; he was a member of the U.S. National Defense Committee and the National Academy of Sciences.

William Shockley joined Bell Labs in 1936, having graduated from the California Institute of Technology and received a Ph.D. degree from the Massachusetts Institute of Technology. He is credited with the invention of the junction transistor—a major step forward that overcame the limitations of the point-contact transistor. His wartime work included antisubmarine research and direction of a major solid-state research program.

In 1955 he established the Shockley Semi-conductor Laboratory at Palo Alto, California, and continued work in the transistor field until 1965. His awards included honorary degrees from the universities of Rutgers and Pennsylvania, and a Professorship of Engineering Science at Stanford University in 1963 which he held until 1975.

Shockley, who died in 1979, was a highly controversial figure in a field far from semiconductor research—no less than control of the ongoing evolution of the human race by genetic manipulation, a theme that occupied him increasingly in the last two decades of his life. His proposals, which aimed at raising the average level of human intelligence by discouraging population increase among those with lower levels of intelligence, were found by many to be totally unacceptable.

THE KEY INVENTION

■ he work of early designers laid a foundation which was waiting for the key invention. The arrival of the key was first announced with this newspaper cutting from the New York Times in 1948. It was given only a small space on an inside page. It announced the discovery of the transistor. From this small announcement our story develops – slowly at first but now at great, and ever increasing, speed. The first transistors were laboratory curiosities quite unsuited to mass production. It was more than 10 years – about 1960 – before production techniques were developed which allowed photographic – like printing processes to be applied to the mass production of transistors and then integrated circuits of greater and greater complexity.

The original point contact transistor made at Bell Laboratories USA.

This process was called the PLANAR process because all of the manufacturing processes were carried out on one side of a flat wafer of material. After this the speed of change increased as mass production led to falling prices and increasing reliability.

The first use of the PLANAR process was to make single transistors – each one on its own part of the silicon wafer. The final

Bardeen, Shockley and Brattain inventors of the transistor.

Figure 12.1. The invention of the transistor (1948), as reported in the *Bell Laboratories Record*, December, 1972. (Left to right) John Bardeen, William Shockley, and Walter Brattain, the inventors.

A proposal that he be awarded an honorary degree at Leeds University in 1973 was withdrawn. He was a singularly difficult person to deal with—he insisted on tape-recording virtually all interviews, and claimed that "the tape-recorder was the single most important application of the transistor."[6] Nevertheless, his contributions to the development of the transistor were profound and far-reaching in their effects.

THE PHYSICS AND TECHNOLOGY
OF THE TRANSISTOR*

To understand a little of what the inventors of the transistor—a term suggested by J. R. Pierce of Bell Labs—had achieved, an outline of the physics and technology involved may be helpful.

The atoms of solids are bonded together by a combination of forces. For the elements silicon and germanium, the four valence electrons (the outermost shell) play a key role. Each atom is sited at the center of a regular tetrahedron formed by its four nearest neighbors, bound to each of them by two coupling electrons—one donated by it and the other by its neighbor. There are no surplus, free, electrons in a pure crystal to conduct electricity when an electric field is applied. If, however, some of the atoms are replaced by atoms of elements in Group 5 (e.g., phosphorus or arsenic), their fifth valence electrons are surplus to bonding requirements and are free to conduct electricity; the material is then said to be "n" (for negative) type. On the other hand, if the replacement is by atoms of elements in Group 3 (e.g., boron or indium), there are insufficient bonding elements and what are called "holes" are created. These holes are also free to move in an electric field (though with lower mobility than the electrons in n-type material) and do so as if they carried a positive charge; the material is said to be "p" (for positive) type. If a piece of pure semiconductor is doped in one region to be n-type and to be p-type elsewhere, the junction between the two will not conduct electricity when the n-region is made positive with respect to the p-region; the potential barrier at the junction cannot be overcome by the barriers on either side. When the polarity is reversed, a current will flow, and both the electrons and the holes crossing will have finite

---

*This summary is based on a contribution from Dr. J. R. Tillman.

lifetimes in regions where normally they do not exist. In an *n-p-n* sandwich structure, with the first *n*-region (the emitter) negatively biased with respect to the *p*-region (the base) and the second region positively biased, we have the junction transistor proposed and analyzed by Shockley, even before he and his colleagues had fabricated one.

The electrons injected into the base region diffuse with little loss to the second junction, which is favorably biased for their immediate collection. If a signal is applied to the low-impedance emitter-base terminals, that derived from the high-impedance collector-base terminals can show very considerable power amplification. A *p-n-p* sandwich will behave similarly, with electrodes biased in the opposite directions to those for *n-p-n*.

Bardeen and Brattain's point-contact transistor in fact used a thin slab of germanium with gold and tungsten contacts, reminiscent of the "cat's whisker" crystal detectors in the early days of radio broadcasting. It demonstrated amplification but lacked stability and was not suited to large-scale manufacture. Shockley's work suggested the replacement of the point contacts by two more layers of doped germanium with different doping from the center slice—and the junction transistor was born. Not only did it offer amplification in a remarkably compact and reproducible device, it could function effectively with low-voltage DC input supplies much smaller than were required by the thermionic valve it replaced.

One of Shockley's earliest ideas had been to try to control the flow of electrons in a semiconductor device by an electric field imposed from outside. Later in the 1950s this possibility of control by an external electric field was reinvestigated and the result was the field-effect transistor—the type most commonly used today.

But equally important as the demonstration of experimental devices that worked and were repeatable was the gradual creation of a viable physical theory that not only explained how transistors worked but also provided a basis for quantitative design and ongoing development.

AND THESE WERE THEIR WORDS

In the commemorative issue of the *Bell Laboratories Record* of December, 1972—"25 Years of Transistors"—the President of the Laboratories and the inventors of the transistor made the following observations.

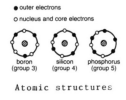

Atomic structures

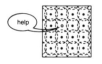

Phosphorus in silicon
(An n-type dopant)

Boron in silicon
(A p-type dopant)

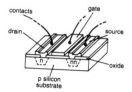

Schematic of an MOS transistor

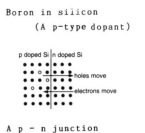

A p - n junction

Figure 12.2. How an MOS junction transistor works. Photograph (opposite) shows section through an MOS transistor.

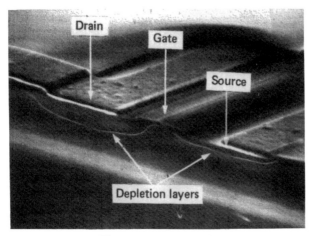

Figure 12.2. (Continued)

James B. Fisk:

> The transistor began with basic research, in Bacon's words, into the "secret motions of things"—the mysteries of solid-state physics—and has led to "the enlarging of the bounds of human empire." But the greatest impact of the transistor may be yet to come—if its benefits can serve as a reminder to society that solutions to many of our problems today, as well as human progress, will depend heavily on the continued pursuit of science and its responsible application.

And this was written before the microchip had revolutionized electronics and its applications in telecommunications, broadcasting, and computing.

John Bardeen:

> They were very exciting days after the invention of the point-contact transistor ... to get a good patent it is necessary to have a good understanding of the basic mechanisms, and there were still questions about just how the holes flowed from the emitter to the collector. How important was the surface barrier layer in the transfer of holes from emitter to collector? Shockley initially suggested the junction transistor structure to help understand the mechanism. Independently, John Shire put emitter and collector points on opposite sides of a thin wafer of germanium, and found that the arrangement worked as a transistor. I can still remember the excitement I felt when I first learned of this discovery, which showed definitely that the holes in the emitter could flow appreciable distances through the bulk of n-type germanium.

This reference to John Shire's vital contribution is interesting and ought to be remembered when discussing the invention of the transistor.
  Walter Brattain:

> The transistor and the development of solid-state electronics were technical off-spring of the revolution in physics that was born with the conception of the principles $E = hv$ and $E = m \times c$ squared by Planck and Einstein at the beginning of this century. Since I am the oldest of the three Nobel Prize winners, I can say it all happened in my life-time.
>
> The use of the transistor of which I am proudest is the small battery-operated radio. This has made it possible for even the most under-privileged peoples to listen ... all peoples can now, within limits, listen to what they wish, independent of what dictatorial leaders might want them to hear and I feel that this will ultimately benefit human society.

William Shockley:

> The creation of solid-state electronics was a complex effort. So many people contributed significantly that I would have to write a book to do them all justice. Instead I have focussed on what might be over-looked—the foundation that supported the entire effort—the managerial skills that created the atmosphere of innovation at Bell Labs. These skills were essential to creating the transistor and to contributing the many benefits that have come to society as consequences.

APPLICATIONS OF TRANSISTORS IN TELECOMMUNICATIONS AND BROADCASTING

During the two decades following their invention in 1948, at first slowly but with increasing tempo, transistors began to have a major impact on the design of telecommunication systems, improving performance and reliability, and reducing size and power consumption. Similar advantages became available in sound and television broadcasting equipment; broadcasting receivers, with mass markets measured in millions, benefited especially from the replacement of valves by transistors. Compact and efficient battery-operated receivers became available at relatively low cost.

The expanding role of transistors and other semiconductor devices in telecommunications has been surveyed by Dr. J. R. Tillman of the British Post Office Research Department in Ref. 2; the following summary is based on that study.

The junction transistor made its first application in telecommunications in transmission systems in the 1950s, then entirely analogue, by effectively replacing thermionic valves. By the 1960s the junction transistor had been developed to a point where inland cable systems carrying more than 10,000 frequency-division multiplex telephony channels on 9.5-mm coaxial cables were possible (Chapter 10); transistors also replaced valves in the baseband and intermediate-frequency sections of microwave radio-relay systems (Chapter 11).

Transistors found important applications in submarine cable systems, beginning in the mid-1960s. The United Kingdom was early in this field with several cables up to 1500 km long, using silicon planar transistors and providing 480 both-way telephone circuits on a single cable. AT&T, using a germanium mesa transistor developed by Bell Labs, laid an 800-circuit cable between the United States and Spain in 1970.

Meanwhile, scientists and engineers at the British Post Office Research Department at Dollis Hill, London, had developed under the direction of Dr. J. R. Tillman high-performance silicon transistors of very high reliability which made possible an 1800-telephone-circuit submarine cable, designed and manufactured by Standard Telephones and Cables (U.K.) and laid between the United Kingdom and Canada in 1973–1974. The reliability standards demanded by such systems with their many repeaters are extremely high and their achievement a demanding scientific and engineering task—typically there must be not more than one failure in 2000 transistors over a period of 25 years.

Much credit for pioneering work in the design, assessment, and manufacture of ultrahigh-reliability transistors for submarine cable systems must go to the British Post Office team responsible for this work: Dr. J. R. Tillman, D. Baker, and M. F. Holmes. They evolved methods for reliability assessment that included scientifically based accelerated aging tests involving very severe overstressing by raising the operating temperature of transistors randomly selected from production batches, followed by electrical tests for any systematic or random changes in performance. (The ability to carry out accelerated aging tests with transistors, not possible with thermionic valves, was a significant factor in moving from valves to transistors in submarine systems.) Their choice of silicon for submarine system transistors, as compared with Bell Labs' choice of germanium, was shown to be fully justifed by the superior performance achieved with silicon.[7] For its work on high-reliability

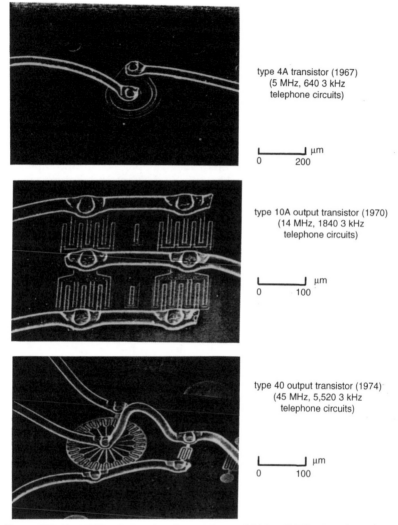

type 4A transistor (1967)
(5 MHz, 640 3 kHz
telephone circuits)

type 10A output transistor (1970)
(14 MHz, 1840 3 kHz
telephone circuits)

type 40 output transistor (1974)
(45 MHz, 5,520 3 kHz
telephone circuits)

Figure 12.3. Post Office-designed and -manufactured high-reliability transistors for submarine cable systems. (Since 1967 more than 10,000 transistors, then valued at at least £5 million, were made and delivered for submarine system use by the PO Research Department at Dollis Hill and Martlesham.)

transistors, the British Post Office Research Department was awarded the Queen's Award to Industry in 1972.

Dennis Baker was awarded the British Telecom Martlesham Medal—a prestigious award made to present or former members of British Telecom staff who have made an outstanding contribution to telecommunications science and engineering—in 1981 for his work in the semiconductor field. His reputation was worldwide: John Bardeen of Bell Labs, coinventor of the transistor, said "it was a well deserved honour," and similar praise came from Ian Ross, President of Bell Labs.

Although the first major applications of transistors in telecommunications were to analogue systems requiring linear amplification over a

Figure 12.4. Award of British Telecom Martlesham Medal to Dennis Baker (1981) for his work on high-reliability transistors for submarine cable systems.

wide range of frequencies, the suitability of the transistor as a low-power-consuming logic and switching element providing fast transition between clearly defined voltage or current levels resulted in important new applications in computers, digital transmission (Chapter 13), and electronic exchange switching systems (Chapter 14). In association with other components, such as diodes and resistors, they formed the basis of logic gates (enabling digital signals to be combined in an "and" gate, selected in an "or" gate, or rejected in a "nor" gate). They also enabled bistable or "flip-flop" circuits, counters, and adders to be made.

In communication satellite systems, transistors and semiconductor diodes provided low-noise microwave amplifiers, and silicon-based solar cells converting sunlight to electricity provided power for satellites (Chapter 15).

But, above all, it was the reliability, small size, low power consumption, economy, and versatility of transistors and other semiconductor devices that provided the essential keys to major advances in the services that telecommunications, broadcasting, and computers could provide.

THE NEXT STEP FORWARD IN THE ELECTRONIC REVOLUTION: THE INVENTION OF THE MICROCHIP

Once it had been demonstrated that individual resistors and capacitors, as well as transistors and diodes, could be created on silicon by suitable doping with boron or phosphorus, it was perhaps inevitable that the idea should arise of forming and interconnecting a number of such components on one slice of silicon to make a "planar" integrated circuit. (The term *planar* indicates that the devices are formed in a single uniform plane).

The need for the integrated circuit first became apparent as the circuitry of military rocket control systems grew increasingly complex and the lack of reliability of conventional printed circuit boards with individually assembled and wired electronic components became more evident and, in some cases, disastrous.

From simple beginnings of just a few components formed on a single slice of silicon to make an oscillator or amplifier, a remarkable technology evolved that ultimately enabled the increasingly complex circuitry of computers, electronic exchange switching systems, and television receivers to be built in fingernail-size microchips.

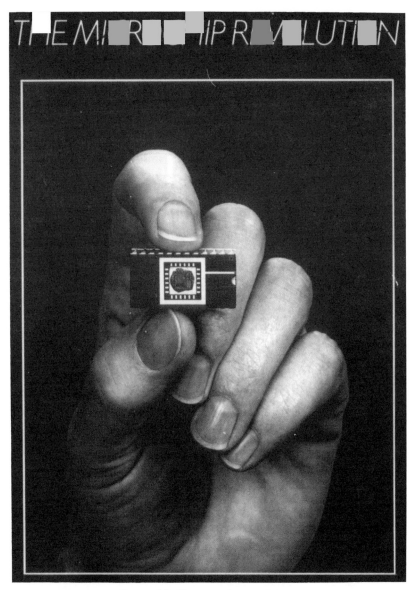

Figure 12.5. The microchip revolution.

Early ideas for an integrated circuit were put forward by G. W. A. Dummer of the U.K. Royal Aircraft Establishment in 1952, with military control systems in mind, but these were not followed up. It remained for two Americans, Jack S. Kilby and Robert N. Noyce, to receive the accolade in October, 1989, of the Draper Award and a prize of $175,000 each from the U.S. National Academy of Engineering for "their separate co-invention of the single-crystal integrated circuit, better known as the semi-conductor micro-chip." The award was also based on "their skill in bringing the integrated circuit to successful production and application in commercial products." An illuminating account of both men and the steps that led them to the invention of the integrated circuit is given by T. R. Reid.[8]

Jack Kilby was born in Jefferson City, Missouri, in 1923; he graduated from the University of Illinois in 1947 and received a Master's degree in electrical engineering from the University of Wisconsin in 1950. He joined Texas Instruments and it was there that his work on the integrated circuit came to fruition in July, 1958. He retired from Texas Instruments in 1970 with more than 60 U.S. patents to his credit. His many honors include the U.S. National Medal of Science, and the U.S. Institute of Electrical and Electronics Engineers Sarnoff and Zworykin medals.

Robert Noyce was born in Burlington, Iowa, in 1927; he received a doctorate in physical electronics from Massachusetts Institute of Technology in 1953. After leaving college he worked for a time in the Shockley

Figure 12.6. The coinventors of the planar integrated circuit—the microchip (1978/ 1979): (left) Robert Noyce (Intel) and (right) Jack Kilby (Texas Inst.).

Semi-conductor Laboratory. He became cofounder of Fairchild Semi-conductor Corporation in 1957, and in 1968 cofounded with G. E. Moore the Intel Corporation with the objective of making large-scale integration a practical reality. He holds 16 patents for semiconductor devices and his many honors include the U.S. National Medal of Science in 1979, the U.S. National Medal of Technology in 1987, and the U.K. Institution of Electrical Engineers Faraday Award.

At Texas Instruments Jack Kilby's first crude attempt at an integrated circuit used a slice of germanium on which were formed a transistor, a capacitor, and three resistors to make a simple phase-shift oscillator. On September 12, 1958, the world's first integrated circuit worked and a new era of electronics was born. However, the components in Kilby's microchip were linked by fine gold wires, and a further important step remained to be taken—the deposition of connections as part of the manufacturing process, a step to which Noyce made a vital contribution.

At Fairchild Semi-conductor in January, 1959, Robert Noyce wrote in his notebook:

> it would be desirable to make multiple devices on a single piece of silicon in order to make connections between the devices as part of the manufacturing process, and thus reduce size, weight etc as well as cost per active element.

And thus the idea of the monolithic integrated circuit was born.

Among their fellow engineers, Kilby and Noyce are referred to as coinventors of the microchip, a term that both men find satisfactory. However, it took several years of legal argument before Texas Instruments and Fairchild came to an agreement to cross-licence their patents, and make them available to other firms on a licensing basis.

THE ONGOING MARCH OF SEMICONDUCTOR DEVICES AND TECHNOLOGY

The years that followed the events of 1958–1959 saw an intensive development of the design, technology, and manufacturing processes of the integrated circuit, leading to dramatic increases in the number of devices that could be accommodated on each chip and corresponding reductions in cost per device.

The number of devices on each chip a few millimeters square has progressed from tens in the 1960s to millions or more in the 1980s, that is, from small-scale, through medium-scale, to large-scale integration, with the prospect of even larger-scale integration to come. This has led to microchips capable of increasingly diverse and complex functions, of high reliability and acceptable cost.

The manufacturing techniques that have evolved over the two decades since the invention of the transistor begin with the production of an ultrapure silicon crystal in rod form which is then sliced into thin wafers. Up to some hundreds of chips are then formed on each wafer by refined photolithographic techniques in which light is projected onto a light-sensitive coating on the wafer through a mask bearing images of chip patterns. The masks are themselves created from a single large drawing of a chip pattern, via a step and repeat camera and photoreduction. After chemical development of the chip patterns on the wafer, the doping of required areas and the formation of conducting connections by metallic deposition is then carried out. Finally the chips are separated from the wafer, mounted on a header, and sealed. These processes are carried out by highly automated manufacturing and testing equipment, under conditions of extreme cleanliness. It is by automation of the manufacturing processes and careful control to produce a high yield of tested and proven chips from each wafer that the cost per chip is kept down.[1,9]

These developments have made possible memory devices capable of holding in digital form the equivalent of tens or hundreds of thousands of words, microprocessors that control the functions of a computer or other electronic control process, e.g., in an electronic telephone exchange switching system (Chapter 14), and even, in the limit, a complete computer on a single chip.

And beyond these levels of integration, still larger-scale integration using electron beams or X rays instead of visible light, is reducing the device size to molecular proportions with many millions of devices per chip. At present the microchip exists only in planar, i.e., two-dimensional, form. It is not inconceivable that it could become three-dimensional, with a further vast increase in capacity—perhaps enabling an hour-long television program to be stored on a single chip.

From the first step in semiconductor technology that began in 1948, many other devices have evolved, some for use at microwave frequencies such as the IMPATT diode, charge-coupled devices (CCDs),

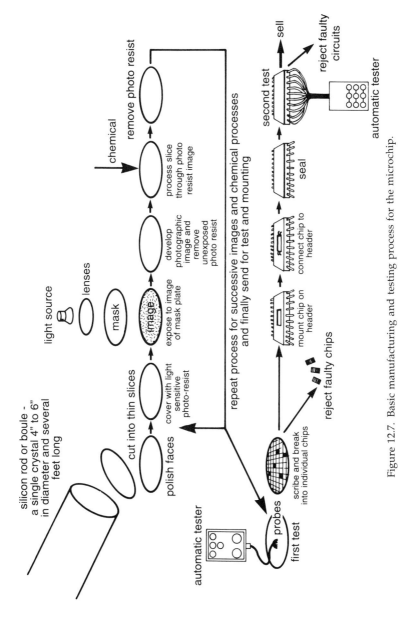

Figure 12.7. Basic manufacturing and testing process for the microchip.

195

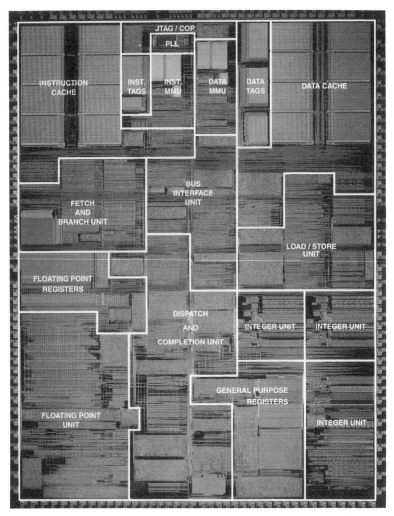

Figure 12.8. Large-scale integrated circuit (microchip) processor for personal computers and word processors. The Motorola PC 604 microprocessor shown contains 3.6 million transistors on a 12.5 × 16-mm silicon chip with 0.65-micron spacing.

and light-emitting diodes (LEDs) that have found application in television cameras and visual displays, and solid-state lasers for optical fiber transmission systems. The year 1948 was indeed the beginning of a major revolution in electronics that transformed telecommunications, broadcasting, and computing.

Germanium, the semiconductor first used to make transistors on any large scale in the 1950s, and by Kilby for the first integrated circuit, soon gave way to silicon for several reasons.

Other semiconductors, based on compounds of elements of Group 3 with those of Group 5 of the periodic table, e.g., gallium arsenide and indium phosphorus, first explored by H. Welker of Siemens in the 1950s, have become important for devices such as lasers, light-emitting and IMPATT diodes.

However, not all inventors of integrated circuit technology accept the priority of Noyce and Kilby. A press report in the U.K. *Sunday Times* in 1990[10] states that a 52-year-old computer expert named Gilbert Hyatt living in Los Angeles, California, had been granted a patent for the invention of a "blueprint" that would enable a computer circuit to be etched on a silicon chip and create the basis for a microprocessor. Hyatt's patent was first registered in 1968, and not granted until 20 years of litigation and expert study of counterclaims had passed. The report also claims that the first commercial microprocessor, made by Ted Hoff at the Intel Corporation, USA, in 1971, was based on concepts originated by Hyatt.

REFERENCES

1. W. H. Mayall, "The Challenge of the Chip," Science Museum, H.M. Stationery Office, 1980.
2. J. R. Tillman, "The Expanding Role of Semi-conductor Devices in Telecommunications," *Radio Electron. Eng., 43,* No. 1/2, January/February 1973.
3. H. E. Stockman, "Oscillating Crystals," letter in *Wireless World,* May 1981.
4. W. Gosling, "The Pre-History of the Transistor," *Radio Electron. Eng., 43,* No. 1/2, January/February 1973.
5. "25 Years of Transistors," *Bell Laboratories Record,* December 1972.
6. Obituary Notice, Prof. William Shockley, *Daily Telegraph,* August 15, 1989.
7. M. F. Holmes, "Active Element in Submerged Repeaters; the First Quarter Century," *Proc. IEE, 123,* No. 10R, October 1976.

8. T. R. Reid, *The Chip: The Micro-Electronics Revolution and the Men Who Made It* (Simon and Schuster, New York, 1985).
9. *The Micro-chip Revolution*, British Telecom Education Service, ES57, 1985.
10. ''Father of the Micro-chip On Line for Millions at Last,'' *Sunday Times*, September 9, 1990.

Further Reading

P. R. Morris, *A History of the World Semiconductor Industry*, IEE History of Technology Series, No. 12, 1990.

13

The Creators of Information Theory, Pulse-Code Modulation, and Digital Techniques

INFORMATION THEORY AS A VITAL KEY TO PROGRESS

The better understanding of the nature of information, e.g., as conveyed from a sender to a receiver by telephonic speech, a facsimile picture, or a television program, and the development of theories providing a quantitative approach to design have been vital for progress in the evolution of more efficient telecommunication transmission and broadcasting systems. The initial, and major, steps forward were the work of a few innovative and farsighted individuals in the laboratories of the American Telephone and Telegraph Company in the United States, Standard Telephones and Cables Ltd. in England, and the International Telephone and Telegraph Company in France.

The existence of the sidebands of a modulated carrier wave had been demonstrated mathematically by J. R. Carson and G. A. Campbell of AT&T Bell Laboratories in 1915 (Chapter 6); their work was extended in the 1920s and 1940s to include the concept of information theory by two other members of the Laboratories, R. V. L. Hartley and C. E. Shannon.[1]

R. V. L. Hartley

R. V. L. Hartley, a Rhodes Scholar, graduated from the University of Utah and later carried out original work on valve circuits—including the invention of an oscillator circuit that bears his name—in AT&T's Western Electric Laboratories. His interest in information theory was stimulated by earlier work by Nyquist of AT&T on the maximum rate at which telegraph signals could be transmitted in a given frequency bandwidth. His studies, published in 1928,[2] resulted in rules for the design of transmission circuits that were in use for more than two decades until C. E. Shannon greatly broadened their scope, and introduced the effects of noise in limiting the rate at which a communication channel could transmit information.

C. E. Shannon: Master of Information Theory

Claude E. Shannon has been described as a mathematician and educator whose information theory profoundly changed scientific perspectives on human communication and greatly facilitated the development of new technology for the transmission and processing of information. It became an important milestone in the transition of society from an industrial base to an information base. His *Mathematical Theory of Communication* was first published as a Bell System Monograph in 1948, and later in book form with Warren Weaver.[3]

He was born in 1916 in Gaylord, Michigan, and graduated with a degree in electrical engineering from the University of Michigan, later receiving M.S. and Ph.D. degrees in mathematics from the Massachusetts Institute of Technology. He joined Bell Laboratories as research mathematician in 1941 and in 1956 became Professor of Communication at MIT where he held the Donner Chair of Science. He received many honors, including the U.S. National Medal of Science and an IEEE Prize Award.

Shannon's Master's thesis, completed at the age of 21, established important relationships between Boolean algebra and the design of switching networks and computers, and remains as a classic in its field. He also made significant contributions to the theory of cryptography as applied to communication security. His early work on cryptography, published in 1945 as a confidential monograph *Communication Theory of Secrecy Systems*, discussed variable coding transformations that made the enciphered text seem random. Some of these ideas emerged later in his information theory which centered on the idea that coding was essential to all communication, i.e., the process of encoding a message into a form suitable for transmission and decoding the message by applying the inverse of the encoding transformation. The performance of a communication channel in the presence of noise could be assessed mathematically from appropriate statistical descriptions of the message information and the noise.

The full development of Shannon's information theory is complex and involves abstruse philosophical concepts as well as sophisticated mathematical analysis.[3-5] However, one result in particular is of direct value to communication engineers. It involves the concept that the smallest unit of information is the "bit," e.g., represented by a simple "on–off" signal, and enables the information-handling capacity C of a communication channel with a signal-to-noise power ratio S/N in a bandwidth B to be given by

$$C = B \log_2 (1 + S/N) \qquad \text{bit/sec}$$

The application of this formula reveals that most practical communication systems have considerable redundancy. For example, a telephone channel of 3-kHz bandwidth and 40-dB (10,000 : 1) signal-to-noise ratio has a theoretical information-handling capacity of 40,000 bits/sec, whereas most practical systems are engineered to handle data at rates that are a fraction of this value. However, the result does give a target against which the efficiency of a given coding system may be measured—an important consideration, for example, in television transmission where the redundancy of practical systems is generally very high, i.e., the efficiency is very low, and the theoretical scope for improvement is considerable.

THE DIGITAL REVOLUTION

A "digital" system is one in which information, whether in the form of speech, facsimile, television, or data signals, is transmitted or processed by "on" and "off" pulses of a current, radio, or light carrier wave, as compared with an "analogue" system in which the continuously varying amplitude of the information signal is conveyed by corresponding amplitude, frequency, or phase modulation of a carrier wave.

Digital techniques, which first began to find application in the 1930s, have now revolutionized telecommunications, broadcasting, and computing by providing superior quality of transmission and recording, greater flexibility in the processing of information, and substantial economic advantages relative to analogue systems. These derive from the basic "on–off" simplicity of digital signals which enables noise and interference to be rejected, nonlinear distortion avoided, and a number of communication channels interleaved in time on a common transmission path without mutual interference.

The principle of time-interleaving was recognized as early as the 1850s in the Baudot multiplex telegraph system which used a mechanical commutator to interleave a number of telegraph channels on a single pair of wires. An attempt was made by Miner in 1903 similarly to interleave speech channels by sampling the speech waveforms with a mechanical commutator at 4000 samples a second—before information theory made it clear that acceptable quality required a sampling rate at least twice the highest speech frequency. It is also relevant that Morse's telegraph (Chapter 3) was essentially a coded digital system.

In the United States P. M. Rainey of AT&T/Bell suggested in 1926 that the elements of a picture might be represented, for transmission purposes, by coded binary, i.e., "on–off," signals to circumvent imperfections in the transmission medium (U.S. Patent No. 1,608,527, November, 1926). In France time-division multiplex methods were studied from 1932 at the laboratories of Le Matériel Téléphonique, where Deloraine and Reeves were granted patent rights on pulse-time and pulse-code modulation methods in 1938.[6]

During the war years the British and American armies used eight-channel pulse-time modulation microwave radio-relay systems, but civil applications lagged until the 1960s when the principles and advantages of pulse-code modulation became better understood and the technology for implementation—notably the transistor and microchip—became available. Meanwhile, frequency-division multiplexing on cables and frequency-modulation radio-relay systems held sway (Chapters 6 and 11).

PULSE-CODE MODULATION: THE KEY INVENTION

Until the advent of pulse-code modulation, digital communication systems had used width or time modulation of individual pulses, each bearing a single voice or other channel. Such systems achieved little, if any, signal-to-noise advantage compared with frequency-modulation systems with the same mean transmitted power. The key invention that conferred on digital systems a virtually complete immunity to noise, interference, and nonlinear distortion was pulse-code modulation—an invention that created a revolution in the transmission, switching, and storage of information.

Pulse-code modulation (PCM) involves sampling the instantaneous amplitude of any analogue signal, whether speech, facsimile, or television, at an appropriate rate and translating each sample into a short train of pulses or "code," with a number of pulses in each train determined by the type of signal and the quality of transmission required. For example, telephone speech is typically sampled at 8000 times per second and each sample is coded into 8 pulses, corresponding to 2 to the power of 8, i.e., 256, discrete amplitude levels of the speech signal waveform, the pulse rate for a single speech channel then being 64,000 bits/sec. On the other hand, a high-quality television signal may require a pulse rate of 100 Mbit/sec. However, more sophisticated coding techniques—embodying some of the concepts foreshadowed by Shannon on the nature of information—can reduce the pulse rate required for television signals substantially.

The noise immunity of digital PCM signals is created by the ability to regenerate the signals, e.g., in repeaters, by circuits that respond only to each pulse amplitude above a predetermined level and reject noise below that level.

The coding of each signal level sample into a group of on–off pulses (called a "byte") is usually carried out in "binary notation." In binary notation a decimal number, which could represent the level of a signal sample, is represented by a succession of on or off pulses in which the first stands for one, the second for two, the third for four, the fourth for eight, and so on. Thus, a sample of level decimal number 11 is represented by binary 1101, i.e., one plus two, no four, and plus eight.

However, digital PCM involves a substantial increase in the frequency bandwidth required for transmission; for example, a 64,000-

bit/sec telephone speech circuit requires a bandwidth, according to Hartley's law, of at least 32 kHz, compared with the basic speech band-width of about 3 kHz. This, together with the lack of suitable device technology, delayed the practical application of digital PCM for two decades or more after its invention in 1937.

The benefits of PCM may be summarized as follows:

• Digitally encoded signals may be transmitted over long distances, requiring many repeaters in tandem, or switched in many ex-changes, without introducing noise, distortion, or loss of signal strength, whereas analogue signals would suffer a progressive loss of quality.

• Many different types of signal, e.g., speech, data, facsimile, and television, can be accommodated on a path such as a pair of wires, a coaxial or optical fiber cable, or a microwave carrier, without mutual interference.

• Digital signals may be stored, e.g., on magnetic tape, disks, or solid-state devices, without loss of quality from repeated use or other causes.

• Digital PCM signals lend themselves to the use of microchips for coding, decoding, regeneration, and information storage, with their advantages of low cost, reliability, and small size.

Alec Harley Reeves (1902–1971): Inventor of PCM

Alec Reeves was born in Redhill, Surrey, England, in 1902. After graduating with a degree in electrical engineering from the City and Guilds Engineering College, London, he joined the staff of International Western Electric (now part of Standard Telephones and Cables Ltd.) and worked on long-wave radio communication across the Atlantic in the 1920s. Over the next 10 years he contributed to the development of shortwave and microwave radio systems. His work in this area, and especially on multiplex telephony, undoubtedly made him aware of the need for a more effective means of reducing noise and crosstalk in such systems.

In his lifetime Reeves was awarded more than 100 patents, but by far the most important and far-reaching in its effect was that of PCM made in 1937 when he was working in the Paris laboratories of Le Matériel Téléphonique. His British PCM patent was granted in 1939.

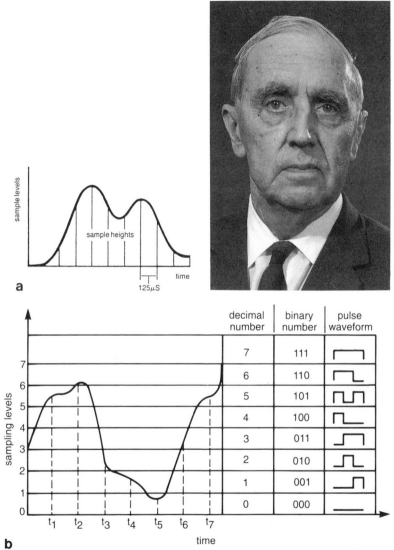

Figure 13.1. Alec Reeves and the invention of pulse-code modulation (1937). (a) Sampling an analogue signal; (b) coding an analogue signal into a binary pulse waveform.

Following the outbreak of the Second World War, Reeves and other colleagues escaped from Paris in June, 1940, to work in the laboratories of Standard Telephones and Cables Ltd. in England. By nature a peace-loving man Reeves was reluctant to work on offensive weapons and his wartime work concentrated on precision bombing aids—including OBOE—which, since it could define military targets with remarkable accuracy, would in the end save civilian lives. For this contribution to the war effort he was awarded the Order of the British Empire.

He was in charge of research on electronic switching systems and devices at Standard Telecommunication Laboratories until his retirement at the age of 69, but continued work in his own research company until his death two years later. This included studies in optical communications on contract to the British Post Office (now British Telecom). It is a tribute to the character of Alec Reeves that only a few days before his death from cancer he was discussing with the author, then Director of Research of the British Post Office, his forward-looking ideas for direct amplification of light in an optical fiber communication system.

A modest and retiring man, Reeves spent much of his private life helping others, including youth and community work and assisting in the rehabilitation of those who had served prison sentences. Only late in his life did the value of his work on PCM come to be understood. In 1969 he was awarded the honor of Commander of the British Empire—and the unique distinction of the issue of a British Post Office stamp commemorating the invention of PCM! The University of Surrey, with the support of Standard Telephones and Cables, has created a Chair of telecommunication studies in his name. He was also awarded the Columbus Medal and the Franklin Medal, and an honorary doctorate of the University of Essex.

In 1990 the Institution of Electrical Engineers, London, created an annual award—the "Alec Reeves Premium Award"—to be given for the best paper on digital coding as judged by the appropriate Professional Group of the Institution—a fitting and continuing tribute to a remarkable man. A colleague of Alec Reeves, Professor K. Cattermole of the University of Essex, has written an illuminating account of the life and work of a man who was one of the most outstanding and innovative scientists of his day.[7]

To quote Professor Cattermole on Alec Reeves:

> He expressed his belief in a better era with "a much saner set of values, based on truth so far as seen, combining science with direct intuitive knowl-

edge" (note his implicit faith that such a combination is possible). Alec Reeves would never have set himself up as an ideal. But his personal integrity, his combination of hope and rationality, of inventiveness and ethics, sets as fine an example as I have ever seen.

THE APPLICATIONS OF PULSE-CODE MODULATION

Perhaps the earliest application of PCM, albeit on a small scale, was in a valve-operated speech security system developed in Bell Labs during World War II for use on shortwave radio links.

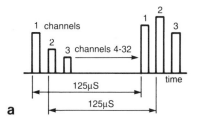

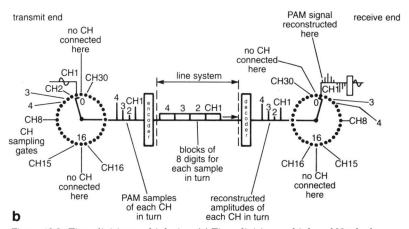

Figure 13.2. Time-division multiplexing. (a) Time-division multiplex of 32 telephone channels (with amplitude modulation of channel pulse samples); (b) principles of time-division multiplex system using pulse-code modulation.

Any large-scale use of PCM, as already noted, had to await the availability of reliable, mass-produced transistors beginning about 1960. The first large-scale application was made in the United States by the Bell System on wire-pair cables providing multichannel voice circuits between telephone exchanges ("junction" circuits), commencing in 1962.[8] This system enabled more channels to be provided on existing cables, with better transmission quality and lower cost. The system—T1 carrier—provided 24 voice channels on a 1.5-Mbit/sec digital carrier with 8-kHz sampling of a 4-kHz voice band and 8-bit logarithmic encoding. Seven bits of the code were used for voice and the eighth for signaling, i.e., for setting up and supervising exchange connections. This built-in signaling feature proved valuable in improving the overall economics of the digital system, as compared with analogue systems using out-of-band signaling.

The overall economics were further improved when, in the 1970s, it became possible to integrate digital transmission on wire-pair junction cables with the new electronic ESS trunk exchange digital switching systems. However, digital transmission on intercity coaxial cables had difficulty in competing economically with the well-established analogue frequency-division multiplex cable systems. Digital microwave radio systems bearing 720 or 1440 voice channels at 45 or 90 Mbit/sec nevertheless proved successful for short interexchange links.

In the United Kingdom the advantages and potential of PCM were early recognized and studies initiated in the 1950s at the Post Office Research Laboratories, Dollis Hill, to optimize the parameters of PCM systems, e.g., sampling frequency, pulses per sample, coding law, signaling methods, and synchronization of pulse trains generated in different parts of the network for voice, television, facsimile, and data signals. In the telephone transmission field, R. O. Carter and Dr. D. L. Richards made notable contributions, as did Dr. T. H. Flowers and W. E. Duerdoth in the digital switching field (Chapter 14).

Although the use of asymmetric-sideband amplitude modulation for television broadcasting had become universal—largely for reasons of bandwidth limitation in the available radio spectrum—the potential of digital techniques for intercity television transmission, the conversion of line-standards, e.g., from U.S. 525 to U.K./Europe 625 lines, the recording of television signals, and the processing of television signals for bandwidth reduction has been both recognized and implemented. The optimization of standards for digital television transmission was also the subject of detailed study at the British Post Office Research Laboratories, in cooperation with the Research Department of the British Broadcasting Corporation and U.K. industry; in this work Dr. N. W. J. Lewis, I. F. Macdiarmid, and J. W. Allnatt at Dollis Hill made important contributions.

These studies, carried out in cooperation with Member Countries and Operating Agencies, led to the recognition—by International Telephone and Telegraph, and Radio Consultative Committees of the International Telecommunication Union—of a "hierarchy" of digital PCM systems in steps ranging from 30 to some 6000 telephone channels at pulse bit rates ranging from 2 Mbit/sec to about 500 Mbit/sec with "slots" for data, facsimile, and television channels. This international cooperation was of great importance in facilitating communication across national boundaries, and in enabling the development of equipment to be carried out in an orderly and efficient manner.

In the United Kingdom the first practical implementation of digital PCM was in 30-channel 2-Mbit/sec systems on wire-pair interexchange junction cables in the middle 1960s, leading to the world's first PCM digital exchange at Empress, London, in 1968 (Chapter 14).

Digital PCM microwave (11 GHz) radio systems providing 2000 voice channels, or a television channel, on each of six carriers at 140 Mbit/sec were also developed and used in the British Post Office trunk network.

A multidisciplinary Task Force set up by the British Post Office in the 1970s to study modernization of the U.K. telecommunications network made a clear recommendation, on economic and operational grounds, to "go digital" for the main intercity trunk network, both for transmission and for switching, and to plan for an "integrated digital network" in which the benefits of digital working could be extended to customers' premises for a wide variety of telecommunication services.

The advent of optical fiber cable with its very wide useful bandwidth, and solid-state lasers that could be readily pulse modulated, opened another door to the deployment of digital PCM techniques (Chapter 17). And satellite communication systems are "going digital" because it offers more effective use of the satellite microwave power, gives greater flexibility in the use of the satellite communication bandwidth, and enables the signals from different earth stations to be readily interleaved in the satellite transponder (Chapter 15).

REFERENCES

1. *A History of Engineering and Science in the Bell System: The Early Years (1875– 1925)*, edited by E. F. O'Neill (Bell Laboratories, 1975), p. 909.
2. R. V. L. Hartley, "Transmission of Information," *Bell Syst. Tech. J.*, July 1928.
3. C. E. Shannon and W. Weaver, *The Mathematical Theory of Communication* (University of Illinois Press, Urbana, 1949).
4. J. R. Pierce, "The Early Days of Information Theory," *IEEE Trans. Inf. Theory, 19*, No. 1, pp. 3–8, 1973.
5. E. C. Cherry, "A History of Information Theory," *Proc. IEE, 98*, No. 3, September 1951.
6. E. M. Deloraine, "Evolution de la Technique des Impulsions Appliqué au Télécommunications," *Onde Electr., 33*, 1954.
7. "The Life and Work of A.H. Reeves," paper presented by Professor K. W. Cattermole at the Institution of Electrical Engineers, London, Discussion Meeting on "Milestones in Telecommunications," November 1989; K. W. Cattermole, "A. H. Reeves: The Man Behind the Engineer," *IEE Rev.*, November 1990.
8. *A History of Engineering and Science in the Bell System: Transmission Technology (1925–1975)*, edited by E. F. O'Neill (Bell Laboratories, 1985).

14

The Pioneers of Electromechanical and Computer-Controlled Electronic Exchange Switching Systems

EXCHANGE SWITCHING SYSTEMS—THE ESSENTIAL LINK BETWEEN YOU AND THE WORLD

With more than 700 million telephones in use worldwide and with facsimile, data, and video communication growing at an ever-increasing rate, the problem remaining once transmission paths have been created by cable, radio, or satellite is to enable any customer to find and communicate with another anywhere in the world. Today this problem has been solved through the efforts of a number of pioneering scientists, engineers, and mathematicians and their successors who have evolved efficient, economic, and reliable exchange switching systems to provide the necessary interconnection between customers, whether in the same town or on opposite sides of the world. The creation of a worldwide telecommunication network has necessarily involved international co-operation on a large scale to secure a commonality of technical and operational procedures to ensure that both local and long-distance connections can be made with equal ease—and which has been achieved through the International Telephone and Telegraph Consultative Committee (CCITT) of the International Telecommunication Union. The world telecommunication network may well be the most extensive, complex, costly, and indispensable artifact yet created—it is in fact the "nervous system" of civilized mankind. The contribution of exchange switching engineers to the creation of this artifact has been critical, demanding both innovative skills and intellectual ability of a high order.

THE ELECTROMECHANICAL ERA
OF EXCHANGE SWITCHING

From the 1900s to the 1970s the automatic exchange switching systems of the world were electromechanical, that is, they used electrically powered relays, stepping mechanisms, or motors to establish connections between an incoming voice channel and a desired outgoing channel selected from banks of tens or hundreds. The history of switching technology is at once complex, detailed, and uses a language all its own; the following brief summary is based mainly on the book by Robert J. Chappuis, the former Head of the Switching and Operations Department of the CCITT in Geneva,[1] *A History of Engineering and Science in the Bell System: The Early Years,*[2] the story of the telephone in the United Kingdom,[3] and papers presented at an Institution of Electrical Engineers (London) Discussion Meeting on "Milestones in Telecommunications."[4]

The first automatic electromechanical system was the Strowger step-by-step system described in Chapter 4, invented in 1889 and which in various modified forms dominated the world's exchanges for nearly a century. The Strowger step system went through a number of stages of evolution at the hands of several inventors, notably by Keith and the Erickson brothers, and Smith and Campbell in the United States from about 1896 to 1915, in order to handle as economically and with as few switches as possible the ever-growing numbers of customers' lines at each exchange.

The basic concepts that made step-by-step a viable switching system were:

1. The 10-point rotary dial which generated pulses of line current according to the digit dialed
2. The two-direction 100-point switch bank which could be controlled by the dialed pulses, and which was also adaptable to various types of connecting function
3. The use of successive switching stages, connected by transfer trunks via 100-point switches
4. The line switch which enabled a given customer's line to be connected to a first selector switch via a free path in a trunk multiple
5. The line finder which looked for an energized line in a multiple of lines and connected it to a trunk and first selector (i.e., the line finder was thus the inverse of the line switch)

It was the application of these concepts that contributed substantially to the economic viability and the quality of service, i.e., reduction of calls failed because of nonavailability of plant, of automatic exchange switching systems throughout the world during the first half of the 20th century, and the principles of which, with different technology, found applications in the electronic switching systems that followed.

Trunking and Grading

An innovation that significantly improved the efficiency of switch utilization was the "graded–multiple" principle invented by E. A. Gray of AT&T in 1905. It involved a departure from the strictly decimal organization of the network of switches in an exchange by allocating say five trunks as first choices feeding perhaps seven selectors, so that fewer selectors are required to handle a given volume of traffic.

Traffic Engineering

Developments such as this inspired a new concept, "traffic engineering," which seeks answers to questions such as "How many calls are to be expected over a given pathway at a given time and what quantity of plant is required to handle these with a defined low probabil-

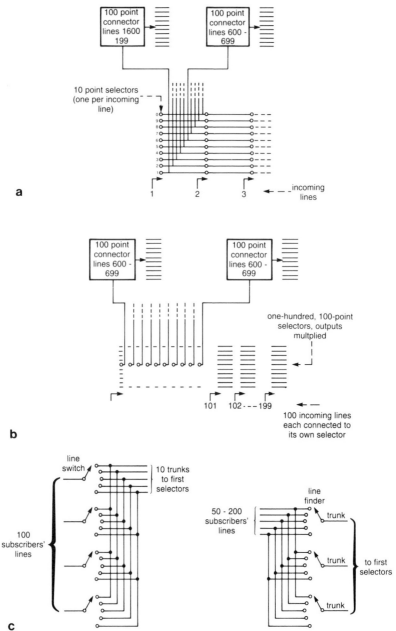

a

b

c

ity of failure?" It involves statistical descriptions of traffic flow, the notion of a "busy hour" in the day when calls are most numerous, and the use of probability theory to assess the percentage of failed calls likely with a given plant configuration. G. T. Blood and M. C. Rorty of AT&T made early contributions to traffic theory, but major steps forward, made in the early 1900s and which remained in almost continuous use since then, were made by E. C. Molina.

E. C. Molina (AT&T): Pioneer of Telephony

From *The Story of the Telephone*[3]:

> Molina was an extraordinary individual who left a lasting mark on the evolution of telephony. He joined Western Electric at the age of 21 with no formal education beyond high school and proved to be a versatile and prolific inventor, making important contributions to machine switching. He joined the Research Department of ATT Co in 1901, and during his long career in this company and with Bell Labs, contributed extensively to the development and application of traffic mathematics. This was made possible by a disciplined self-study of mathematics that led him to become an expert in certain fields, particularly in the work of Laplace. In recognition of this work, and the encouragement it gave to young people for self-study, he received in 1952 the Honorary Degree of Doctor of Science from Newark College of Engineering, an institution at which he taught mathematics until 1964, some 20 years after his "retirement" from the Bell System. [p. 540]

His work also included significant contributions to the development of exchange common-control systems described below.

Traffic engineering studies had also received attention in Europe, notably in Denmark by A. K. Erlang.

A. K. Erlang: Pioneer of Traffic Theory

Erlang initiated a new approach to traffic theory by introducing an exponential distribution of call holding time, compared with the constant average holding time assumed in earlier studies. This enabled greater accuracy to be achieved in the prediction of call failures, and permitted analytic solutions to traffic flow problems not previously

◄————————————————————————————

Figure 14.1. Principles of exchange switching systems. (a) 1000-line system with transfer trunks and 10-point selectors; (b) 1000-line system with transfer trunks and 100-point selectors; (c) line switch (left) and line finder (right).

possible; it also provided a basis for modern traffic "queuing" theory. However, Erlang's work received little attention outside his own country until its publication in the British Post Office *Electrical Engineers Journal* (London) in 1918.

Exchange Switch Development

Although the 100-point (10 by 10) Strowger switch remained in the lead in the world's exchange switching systems for several decades after 1900, a variety of other switch types were developed in the United States and Europe, stimulated by the need to keep down the cost of exchange equipment as the number of customers' lines at each exchange increased into the tens and hundreds of thousands. Typical of these were the linear "panel" switch used by AT&T in the United States, which simplified trunk multiple wiring, and the "rotary" switch used in France and elsewhere in Europe. Switch points increased to 200 or more and electric motor drives were used to power the switches in place of direct control by dialed pulses of current from the customers' lines. "Crossbar switches," pioneered in Sweden and which used plane arrays of rectangularly coordinated contacts, were also developed and widely used.

The problems of switch point contact reliability were tackled, for example in the United Kingdom and the United States, by "reed relay" exchanges in which the contacts were contained in glass tubes sealed against the atmosphere and operated by externally applied magnetic fields.

However, it became increasingly clear that switch development alone did not adequately cope with the problems created by telephone traffic growth—especially evident in big cities such as New York and London. A new approach was required to deal with this situation.

The "Big City" Problem: The Evolution of Common Control
of Exchange Switching

The heart of the "big city" problem was that the volume of exchange switching equipment using step-by-step direct control—despite switch development—increased more rapidly than the number of customers' lines so that a point was reached where such systems became uneconomic and unwieldy beyond a few thousand lines. The solution lay in what came to be known as "common control," a principle that was later applied in computer-controlled electronic exchanges as well as electromechanical ones.

In a common-control switching system the pulses created by the dialing of a call are placed in a temporary store or register and then converted into a machine language best adapted to control the interconnecting switches. The stored signals, in this converted form and under the direction of a built-in program, then control the switches needed to make the desired connection, initiate the process of alerting the called customer, and inform the calling customer of the progress of the call.

Common control not only considerably reduces the amount of switching equipment and the time needed to set up connections, it also provides greater flexibility in handling other types of traffic as well as telephone calls, and the control of fault conditions, by appropriate design of the stored program. It thus embodies many of the features of computers, e.g., in the use of memory, translation, stored program, and machine language, all of which were later achieved in microelectronic, as compared with electromechanical, format.

The London Director System: Col. Purves and G. F. Odell

Following the First World War the British Post Office initiated the progressive introduction of automatic telephone exchanges throughout the country, based largely on the Strowger step system. It was early recognized that London presented a "big city" problem in common with that encountered in the United States. In 1919 consultations took place between BPO staff headed by Col. Purves, then Engineer-in-Chief, and AT&T/Western Electric which led to a decision in 1922 to develop a London Director system of automatic telephony based on the common-control register/translator principles, using Strowger step switches and incorporating the combined letter/numeral numbering system first proposed by W. G. Blauvert of AT&T.[2,3]

The first London Director system installation was made in 1927. G. F. Odell, who was then BPO Assistant Engineer-in-Chief, described it as a "high water-mark in the tide of human creative intelligence." One wonders how he would have described the vastly more complex and sophisticated computer-controlled electronic switching systems that came into being half a century later!

THE COMPUTER-CONTROLLED ELECTRONIC
EXCHANGE SWITCHING ERA

The evolution of exchange switching from electromechanical systems to computer-controlled electronic systems opened the door to major improvements in the speed and reliability of the connection process, but above all it enabled additional service facilities such as abbreviated dialing, call transfer, and itemized billing of call charges to be provided rapidly and efficiently. It greatly facilitated the management of switching exchanges and customers' lines, especially with regard to fault location and correction. And it provided the flexibility needed to handle the expanding nonvoice traffic such as telex, data, facsimile, and video.

This evolution was made possible mainly by the following advances in device technology and system concepts:

- The invention of the transistor and the microchip (Chapter 12)
- The principle of "stored-program control," adapted from computer systems
- The use of digital techniques (Chapter 13)

The transistor and the microchip, with other solid-state devices, provided the basis for fast and very reliable switches, in contrast to slow-moving electromechanical switches which required much maintenance. These devices enabled temporary and semipermanent "memories" to be made for storing information, e.g., in transient form about call progress and, in more permanent form, the programs needed for switch control. They also enabled "logic" operations, e.g., the manipulation of information into a different form suitable for switch control, to be carried out extremely rapidly. In contrast to earlier "wired logic," solid-state logic could be readily rearranged to modify or change the stored program to accommodate changing or new service requirements.

Many of the first electronic exchanges used "space-division" techniques, i.e., each switch handled one call at a time. With the increasing use of digital transmission in junction and trunk networks (Chapter 13), it became desirable and economic to switch on a "time-division" basis, each switch handling many calls interleaved in time—made possible by the fast operating speeds possible with solid-state devices.

Progress from electromechanical to computer-controlled electronic exchanges proved to be a complex process, involving a massive development effort and a (sometimes painful) "learning process" in which early experimental designs had to be subjected to the rigors of field trials on a sufficient scale to reveal any limitations.

Furthermore, technology was advancing more rapidly than development could be completed, notably in the following areas:

- The change from "reed" to digital switching
- The progress of storage technology from magnetic cores to ever more complex semiconductor stores
- The evolution of reliable software
- The need for ever larger scale integration of semiconductor devices on chips to solve difficult design problems cost-effectively, especially analogue-to-digital conversion on a per-customer basis

The following summary draws mainly on experience with early electronic exchange design and field trials in the British Post Office (BPO) and the U.K. telecommunications industry, and in the Bell System in the United States, beginning in the 1960s. Because of the complexity of the work involved, these developments were very much team efforts, and any names mentioned as "pioneers" are essentially those of some of the team leaders. It must also be remembered that similar developments were taking place in other countries, including France, Germany, Sweden, and Japan.[5]

ELECTRONIC EXCHANGE DEVELOPMENT IN THE UNITED KINGDOM

Following the Second World War the BPO Research Department at Dollis Hill, London, recognizing the potential advantages of electronic switching, began experimental studies, concentrating mainly on a multiplexed approach with time-division pulse-amplitude modulation (PAM-TDM) of the speech path as the most likely to yield an economic solution. Major activities in the mid-1950s included the design of electronic register-translators for the London Director and similar switching systems, and magnetic-drum translators for the Subscriber Trunk Dialling (STD) system. These developments benefited considerably from experience gained during the war years in the design by BPO Research staff of the "Colossus" digital computer for cracking German military codes—probably the first full-scale digital computer ever made.[6]

The developments in electronic switching systems technology in the United Kingdom during the period from 1950 to 1975 are surveyed in depth in Ref. 7; only the main highlights are described below.

The Highgate Wood Field Trial

The BPO and five leading U.K. manufacturers of exchange equipment agreed in 1956 to pool research effort in electronic switching in a Joint Electronic Research Committee (JERC). After examining various space- and time-division possibilities, and taking into account the available technology, it was decided to proceed with the design of a PAM-TDM system for an exchange capable of serving up to 20,000 lines. The then available electronic devices included early transistors, diodes, valves, and cold-cathode tubes as switches; magnetic drums were available for code translation and recording class of service, call, and line status information.

The field trial at Highgate Wood, NW London, was commissioned in 1962 with 600 lines; however, by mid-1963 it was concluded that it did not provide the basis of a viable and economic design and was withdrawn from service. The main problems that arose were centered on excessive power consumption, high cost of components, inadequate noise and crosstalk performance, and incompatibility of signaling with the existing network. In retrospect the Highgate Wood field trial can be seen to have been overambitious, bearing in mind the limitations of device technology at the time. But many valuable lessons were learned and the TDM approach was referred back to research, later to emerge in the form of PCM-TDM switching systems.

It is perhaps significant that the Bell System also went through a "learning experience" with its first electronic exchange at Morris, Illinois, when little of the technology used survived in later designs.

Meanwhile, the BPO and the manufacturers concentrated on a range of space-division common-control exchanges (TXE2 and TXE4) using reed-relay switches and well-tried transistor–transistor logic circuitry. These were not strictly electronic exchanges but their use on a large scale allowed time for the development of PCM-TDM switching and System X described below.

Experience with early models of reed-relay exchanges showed that the reed relays, although sealed against deterioration of the contacts by the external atmosphere, nevertheless were not immune to contact failures and required great care in manufacture and application to ensure high reliability. The objective of wholly electronic exchanges remained and was pursued with vigor.

The First Digital Switching Exchange: Empress, London (1968)

The success of pulse-code modulation time-division (PCM-TDM) transmission over junction and trunk routes (Chapter 13) pointed to the desirability, for economic and operational reasons, for switching in tandem and trunk exchanges without demultiplexing to voice frequencies. On the other hand, the relatively high cost of "codecs" for coding voice signals into PCM format and decoding, prohibited the use of PCM-TDM for local exchanges until such time as more cost-effective device technology, e.g., the microchip, became available.

Design studies of a PCM-TDM switching system suitable for a field trial began at the BPO Research Department, Dollis Hill, London, in 1965. The trial system switched traffic on six 24-channel PCM links between three neighboring London exchanges.

It used a space–time–space switching configuration, with diode-transistor logic and some early designs of TTL integrated circuits. The switch paths were controlled by a cyclically addressed ferrite-core store.[8,9]

The BPO field trial system was installed at Empress Exchange, West London, and was opened for live traffic by the then Post-Master General John Stonehouse in September, 1968. It was followed by PCM-TDM tandem switching system, similar to Empress but with processor control of switching, built by Standard Telephones and Cables Ltd. and installed at Moorgate Exchange, London, in 1971.

The success of these digital electronic exchanges led to an intensive study of the main U.K. network by a BPO Trunk Task Force which concluded in 1971 that integrated digital switching and transmission could offer significant economic advantages, which would increase as large-scale integrated circuits became available.

The experience gained with the Empress and Moorgate digital exchanges, and with the evolutionary modular design of the TXE2 and TXE4 exchanges, provided basic inputs to the next major step forward in exchange switching system design—System X.

The Evolutionary Approach to Exchange Switching: System X

The evolutionary System X concept, initiated by the BPO and subsequently developed in collaboration with U.K. industry, defined a com-

Figure 14.2. The Right Honorable John Stonehouse, Post-Master General, opening the PCM–digital exchange at Empress, London, with the author (1968).

mon framework for a family of exchange switching systems including local, tandem, and trunk exchanges, based on a series of hardware and software subsystems each carrying out or controlling specific functions in the switching process. In this way provision was made for future changes, e.g., to take advantage of new system concepts and advances in device technology, and to provide new services rapidly and economically, with a minimum of equipment replacement. At the same time, System X was designed to work with the existing network on either an "overlay" or replacement basis so that the network could be progressively updated without disruption.

Figure 14.3. PCM–digital telephone exchange at Empress, London. Opened in 1968, it was the world's first digital exchange to go into public service.

Central to the System X design were:

- Stored program control based on data processing and computer techniques providing a powerful and flexible means for controlling the operation of the exchange and performing administrative functions such as maintenance, recording call charges, and collecting traffic data
- Common channel signaling which uses one out of perhaps hundreds of traffic channels for providing all of the signaling functions needed to control the setting up and routing of calls

Figure 14.4. System X (a) exchange equipment and (b) digital subunit.

- Provision for integrated digital transmission and switching via local as well as tandem and trunk exchanges
- The use of semiconductor technology, including "microchip" large-scale integrated circuits, e.g., as "processors" for the common control of switching
- Provision for nonvoice services such as telex, facsimile, circuit-switched data, Prestel, and audio and video conference facilities

Many of the System X concepts were initiated in the BPO Research Department at Dollis Hill, NW London, and were later given specific shape by a Joint PO–U.K. industry Advisory Group on System Defini-

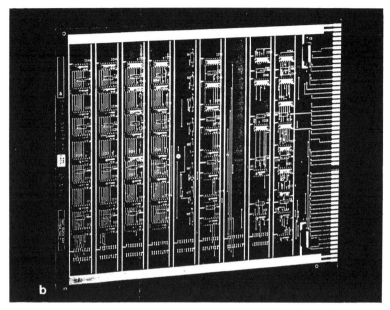

Figure 14.4. (Continued)

tion (AGSD) set up in 1968 to define the architecture of System X and its subsystems.

The subsequent collaborative program for the design and manufacture of System X, involving the British Post Office (later British Telecom), General Electric Telecommunications Ltd., Plessey Telecommunications, and until 1982 Standard Telephones and Cables Ltd., was initiated in 1976 with the aim—subsequently achieved—of large-scale production for the U.K. and world markets.

British Post Office/Telecom Pioneers in Electronic Switching

Any development as complex and technologically challenging as the evolution of electronic switching inevitably involves many creative minds and many contributors to the ultimate outcome. The following

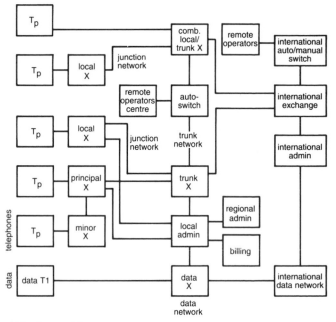

Figure 14.5. Types of System X exchanges (providing local, trunk, and international exchanges, auto-manual and data switching).

individuals have been singled out by "peer" judgment as having made particularly noteworthy contributions.

Dr. Thomas Flowers

Dr. Flowers was responsible for much of the early work in the BPO Research Department at Dollis Hill, NW London, on time-division electronic switching that led to the Highgate Wood field trial. He is well known for his work during the war years on "Colossus"—probably the first digital electronic computer—which achieved major success in breaking German military communication codes. A number of the ideas inherent in Colossus, e.g., stored-program control and memory systems, later found application in electronic switching. He was the first

Figure 14.6. British Telecom Martlesham Medal winners for innovative work in electronic switching: (top) Dr. T. Flowers (1980), (bottom left) Mr. C. J. Hughes (1984), and (bottom right) Mr. L. R. F. Harris and Mr. J. Martin (1983).

to be awarded the Martlesham Medal, in 1980, in recognition of his achievements.

L. R. F. Harris

Roy Harris and John Martin (see below) were jointly awarded the Martlesham Medal in 1983 for their work in masterminding for British Telecom the design and development of System X.

Roy Harris, the son of a former Director of Research and Engineer-in-Chief of the Post Office, had worked with Dr. Flowers on early electronic switching research at Dollis Hill. He led the team that developed the early versions of the reed-relay stored-program common-control exchanges that later became the TXE2 and TXE4 exchanges widely used in the United Kingdom.

He made a vital contribution to System X with the concept of modular equipment, set within an enduring overall system "architecture," with the functions of the modules or subsystems, and the interfaces between them, precisely defined.

Roy Harris was born in 1926; he graduated with first-class honors in Mechanical Sciences at Cambridge University and joined the Post Office in 1947.

He became Director of Telecommunications Systems Strategy in 1974 and has played a leading role in national and European forums aiming at harmonizing development in telecommunications and information technology. There are some 20 patents bearing his name and he has published extensively on switching stategy.

John Martin

John Martin arrived in the System X team at the start of collaborative development with U.K. industry in the 1970s at a time when software-controlled digital microelectronic technology was in its infancy. He became Director of System X Development and was responsible for overall management and leadership of the large joint PO–U.K. industry teams concerned with the development of System X.

He was born in 1927 and joined the Post Office Engineering Department in 1943 as a Youth-in-Training. He obtained a degree by part-time study at Northampton Polytechnic (London). His early work was on the introduction of subscriber trunk dialing in the United Kingdom and development of reed-relay electronic exchange switching systems (TXE3 and TXE4).

In presenting the Martlesham Medal to Roy Harris and John Martin, John Alvey, British Telecom Managing Director (Development and Procurement), said:

> The successful development and supply of System X is the most significant technical event in British Telecom's recent history. The project is the backbone of our digital modernisation; on it ride all the new services which will help to transform the way in which Britain's commercial community goes about its daily tasks.

Charles Hughes

Charles Hughes, who received the Martlesham Medal in 1984, originated the idea of using for telecommunications what the world now knows as the "microprocessor"—a vital component in the design of stored-program control switching systems.

His early work at the BPO Research Department, Dollis Hill, was in the management of the Empress Exchange digital exchange field trial and to which, in collaboration with colleague Winston Duerdoth, he contributed important ideas in the use of PCM-TDM techniques.

Charles Hughes's achievements include:

- The microprocessor concept
- Establishing important principles in the architecture of stored-program control switching systems
- Contributions to the development of digital switching and recorded voice systems
- The concept of a universal digital system of variable capacity accommodating instantaneous demands from a number of constantly varying sources

Charles Hughes was born in 1925; he graduated from London University and received a Master's degree. He joined the BPO as an executive engineer in 1955 after working for a time with Cable and Wireless Ltd., working initially on radio research and then for many years on electronic switching. He became Deputy Director (Switching and Trunk Transmission) in 1975, and in 1987 was awarded a Fellowship of Engineering (London).

ELECTRONIC EXCHANGE DEVELOPMENT
IN THE BELL SYSTEM (USA)

The scientists, mathematicians, and engineers of the Bell Laboratories must be regarded as pioneers of electronic exchange switching based on stored-program common control—their work in the early 1960s and onwards was followed with great interest by telephone engineers throughout the world. It stimulated similar switching developments in many countries, notably in the United Kingdom, Sweden, Germany, France, and Japan.[5] A lively, perceptive, and entertaining account of switching developments in Sweden appears in J. Meurling and R. Jeans's book *A Switch in Time*.[10]

The proving of major advances such as stored-program control demands a field trial on a sufficient scale to demonstrate both practicability and economic viability; for the Bell System, such a field trial took place at Morris, Illinois, from 1960 to 1962.[11]

The Morris, Illinois, Electronic Exchange Field Trial

The design of the experimental exchange at Morris, with up to 400 customers' lines, was based on the use of electronic rather than electromechanical techniques to control and carry out switching operations, enabling millions of individual functions to be carried out in the 10 seconds or so taken to dial a seven-digit number.

It was essentially a "space-division" system using some 20,000 gas-filled tubes as switches, together with customer line scanners and a stored-program common control using a flying-spot storage tube and photographic plates to provide a long-term memory. An electrostatic "barrier-grid" storage tube provided a short-term memory for dialed number and call progress information.

It is of interest that the use of gas tubes as switches with their limited current-handling capability made it necessary to use tone-frequency signaling. Other advanced features in the trial were abbreviated code dialing and call transfer facilities.

Although little of the technology used in the Morris field trial survived in later electronic exchange development, the value of stored-program control, with its demonstrated operational flexibility, was fully proven and adopted in the systems that followed. The faster setting up of calls available with "touch-tone" dialing was appreciated by customers and created an incentive to wider use of this facility.

The experience gained from the Morris field trial—as with the BPO Highgate Wood and Empress field trials—provided both a stimulus and basis for further electronic exchange development, notably in the Bell System with their "No. 1 ESS." A detailed account of this development is given in a special issue of the *Bell Laboratories Record* for June, 1965,[12] on which the following summary is based.

The Bell System No. 1 ESS: A Pioneering and Major Advance in Electronic Switching Systems

The opening of the No. 1 Electronic Switching Exchange at Succasunna, New Jersey, in May, 1965, has been described as "the culmination of the largest single development project ever undertaken by the Bell Laboratories for the Bell System" (W. Keister in Ref. 12).

The basic concept of No. 1 ESS is of a single high-speed electronic central processor operating with a stored program to control the switching operations on a time-sharing basis. Through the switching network, interconnections can be made between customers' lines and trunks, and access provided to services needed to handle calls such as tones, signaling detectors, and ringing. The central processor includes temporary and semipermanent memories, the former for storing transient information required in the processing of calls dialed by the customer, while the latter stores the program for control of the switching operations.

Input and output information for the central processor is provided by scanners, e.g., of calling customers' lines, and distributors, e.g., of signals initiating action required by the central processor.

Although wired logic is used in the central processor, changes in the program can be readily made in the program memory to accommodate a variety of services. The device technology used in ESS 1 included:

- The "ferrod," a magnetic current sensing device associated with each customer's line
- The "ferreed" switch, a two-state magnetic reed switch sealed in a glass envelope
- The "twistor" semipermanent memory, a magnetic sheet capable of storing several million bits of information in magnetized dot

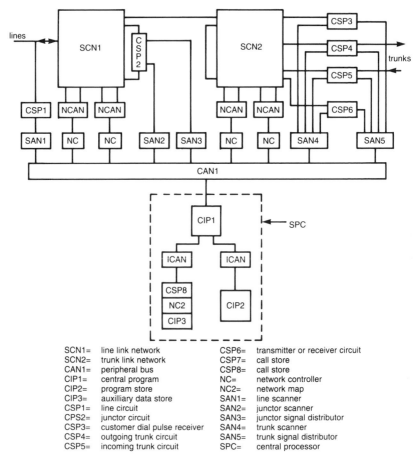

SCN1=   line link network
SCN2=   trunk link network
CAN1=   peripheral bus
CIP1=   central program
CIP2=   program store
CIP3=   auxilliary data store
CSP1=   line circuit
CPS2=   junctor circuit
CSP3=   customer dial pulse receiver
CSP4=   outgoing trunk circuit
CSP5=   incoming trunk circuit

CSP6=   transmitter or receiver circuit
CSP7=   call store
CSP8=   call store
NC=   network controller
NC2=   network map
SAN1=   line scanner
SAN2=   junctor scanner
SAN3=   junctor signal distributor
SAN4=   trunk scanner
SAN5=   trunk signal distributor
SPC=   central processor

Figure 14.7. The first electronic switching exchange, Bell System, USA (1965).

form, the pattern of stored information being alterable by replacing a card between the magnetic memory planes

Thus, the ESS 1 system made effective use of a variety of magnetic
device technology, in addition to silicon transistors in its logic circuits.
In this sense it was perhaps less "electronic" than its successors a decade
or so later which used large-scale integrated semiconductor circuits to

carry out similar functions even more efficiently, economically, and compactly.

In reviewing progress in 1969 during an interview on "ESS—Four Years On,"[13] R. W. Ketchledge, then Executive Director of Bell Labs' Electronic Switching Division, noted that by that date more than 80 ESS systems were in regular service in the Bell System, and observed that

> The important thing about ESS is stored-program control . . . we have enough experience now to know we were absolutely right in 1955 when we gambled on this—it was a big gamble then—and turned the entire development around from a wired-logic system to a stored program system. We could see that stored program could give us a kind of flexibility that switching never had before . . . by changing programs we can add new services more quickly and at lower cost than would be possible by changing hardware.

Bell Pioneers and the Onward March of Progress in Electronic Switching

It is perhaps invidious to try to single out "pioneers" and "innovators" in an operation as complex and challenging as the early develop-

Figure 14.8. R. W. Ketchledge, Director of Bell Electronic Switching Lab (1959). (Author of *From Morris to Succasunna ESS 1*.)

ment of electronic switching in the Bell System—so many people were involved not only in the evolution of new system concepts and the creation of new technology, but also in the innovative management of a huge project, with its involvement of manufacturing on a massive scale by Western Electric and field installation and testing throughout the United States.

An early precursor of ESS 1 is worthy of mention for its pioneering content—the field trial of a stored-program, digital exchange (ESSEX), by Earl Vaughan of BTL.

Some of the contributors to ESS 1 development as recorded in Ref. 12 are:

- W. Keister: "The Evolution of Telephone Switching"
- R. W. Ketchledge: *From Morris to Succasunna*
- Features and Service: J. J. Yostpile
- The Stored Program: E. H. Siegel, Jr., and S. Silber
- The Control Unit: A. H. Doblmaier
- Memory Devices: R. H. Meinken and L. W. Stammerjohn
- The Switching Network: A. Feiner

From the pioneering development of ESS 1, other developments in electronic switching followed in the Bell System, including ESS 2 and ESS 3 for the smaller exchanges in suburban and rural communities.[5]

These, like ESS1, were essentially space-division switching systems. An early exploitation of the possibilities of time-division switching was the 101 ESS offered in 1963 as a private branch exchange for business users.[14]

By the 1970s PCM–digital transmission was beginning to supplement FDM–analogue transmission over junction and trunk routes in the United States as in the United Kingdom and it became operationally and economically desirable to switch in this mode. The outcome in the Bell System was the ESS 4 PCM–digital electronic switching system, designed as a successor to electromechanical crossbar switching systems, to cater economically and efficiently to the massive growth in cross-country telecommunication traffic then beginning.[15]

It is perhaps relevant to note, however, that progress toward ever more complex, highly centralized control of nationwide switching systems, dependent on computer control involving sophisticated software, is not without its problems. Difficult-to-detect software errors have already led to large-scale network "shutdown" possibilities, as has occurred in the United States and elsewhere.

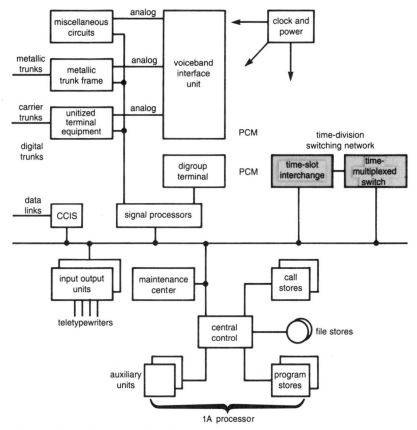

Figure 14.9. Bell System (USA) No. 4 electronic switching system. The No. 4 ESS routes incoming digital signals to the appropriate outgoing trunks under the control of a local processor. Incoming analogue signals on wire or carrier circuits are first converted to PCM–digital format.

REFERENCES

1. Robert J. Chappuis, *100 Years of Telephone Switching: 1878–1978* (North-Holland, Amsterdam, 1982).
2. *A History of Engineering and Science in the Bell System: The Early Years (1875–1925)* (Bell Laboratories, 1975).
3. J. H. Robertson, *The Story of the Telephone* (Sir Isaac Pitman and Sons, London, 1947).

4. Papers presented at The Institution of Electrical Engineers (London), Science, Education and Technology Discussion Meeting on "Milestones in Telecommunications," November 1989. (a) A. Emmerson, "Strowger and the Invention of the Automatic Telephone Exchange (1879–1912)"; (b) S. Welch, "The Early Years in the UK Telephone Service Environment"; (c) W. Duerdoth, "The Introduction of Digital Switching in the 1960s."

5. *Electronic Switching: Central Office Systems of the World*, edited by Amos E. Joel, Jr. (Institute of Electrical and Electronics Engineers, New York, 1976).

6. A. W. M. Coombs, "Colossus and the History of Computing," *Post Off. Electr. Eng. J.*, *70*, July 1977.

7. C. J. Hughes and T. E. Longden, "Developments in Switching System Technology," *Post Off. Electr. Eng. J.*, *74*, Pt. 3, October 1981.

8. K. J. Chapman and C. J. Hughes, "A Field Trial of an Experimental Pulse-code Modulation Tandem Exchange," *Post Off. Electr. Eng. J.*, *61*, October 1968.

9. E. Walker and W. T. Duerdoth, "Trunking and Traffic Principles of a PCM Exchange," *Proc. Inst. Electr. Eng.*, *111*, December 1964.

10. J. Meurling and R. Jeans, *A Switch in Time* (Telephony Publishing Corp. USA, 1985).

11. E. G. Neumiller, "The Electronic Central Office at Morris, Illinois," *Bell Lab. Rec.*, December 1962.

12. "The No. 1 Electronic Switching System," *Bell Lab. Rec.*, June 1965.

13. R. W. Ketchledge, "ESS—Four Years On," *Bell Lab. Rec.*, August 1969.

14. "101 ESS—A More Flexible Telephone Service for Business," *Bell Lab. Rec.*, February 1963.

15. C. D. Johnson, "No. 4 ESS—Long-Distance Switching for the Future," Ref. 5, p. 38, and *Bell Lab. Rec.*, September 1973.

15

The First Satellite
Communication Engineers

THE IMPACT OF SATELLITES ON WORLDWIDE
TELECOMMUNICATIONS AND BROADCASTING

The ability of Earth-orbiting satellites to provide, economically and reliably, direct high-quality communication between any two locations on the Earth's surface, and direct broadcasting coverage over large areas, has expanded enormously the scope and volume of these services.

The technical advantages were obvious enough. For telecommunication links only one repeater was required, i.e., in the satellite, compared with the many repeaters on long land and submarine cables, and land microwave radio-relay systems, with substantial gains in reliability. For direct broadcasting to homes large areas could be covered by a single transmitter in the satellite, without encountering the "shadow" areas created by hills and mountains when using ground-based transmitters. Furthermore, the microwave frequency band most suited to satellite communication systems was immune to the vagaries of the ionosphere and was virtually free from atmospheric absorption over a range from about 1000 to 10,000 MHz, above which rain absorption became progressively more apparent. By sharing frequency usage with compatible terrestrial microwave systems, wide frequency bands could be made available for satellite system use. This factor, coupled with the use of highly directional aerials at earth stations, meant that there was scope for a massive growth of telecommunication traffic on a worldwide scale, involving earth stations ranging from small units providing some tens of telephone circuits to large units with several thousands. Transmission over satellite microwave links proved to be virtually free from the fading and reflection phenomena encountered at times on terrestrial microwave links, and was suitable for high-definition television signals.

Long-distance telecommunication links by satellite also offered operational advantages in that these could bypass intervening countries, avoiding the troublesome problems of wayleaves and transit charges that terrestrial links were prone to. Because of the relatively high initial costs involved, notably for launching satellites, the economics of satellite links compared with land and submarine links are more favorable for the longer distances. And for direct television broadcasting the initial cost per viewer by satellite improves compared with ground-based transmitters as the coverage and the viewing audience increases.

But until the mid-1950s the possibility of communication by satellite was considered by most to be in the realm of science fiction.

EARLY PROPOSALS FOR COMMUNICATION SATELLITES

The first proposals for a communication satellite system were made in 1945 by A. C. Clarke (United Kingdom). Clarke worked initially at the Royal Aircraft Establishment, Farnborough, England, where he doubtless became familiar with the basics of rocketry and satellite orbits. He was later to become famous as a science fiction writer and adviser to the makers of futuristic films such as *2001*. Clarke's proposals, made 12 years before the launching of the first satellite, appeared in a modest article in the journal *Wireless World* in October, 1945.[1] It is perhaps relevant that England, for the previous year or so, had been on the receiving end of German V2 rocket-propelled missiles, and this may have inspired the idea that there were more useful things that could be done with rockets!

EXTRA-TERRESTRIAL RELAYS

Can Rocket Stations Give World-wide Radio Coverage?

By ARTHUR C. CLARKE

ALTHOUGH it is possible, by a suitable choice of frequencies and routes, to provide telephony circuits between any two points or regions of the earth for a large part of the time, long-distance communication is greatly hampered by the peculiarities of the ionosphere, and there are even occasions when it may be impossible. A true broadcast service, giving constant field strength at all times over the whole globe would be invaluable, not to say indispensable, in a world society.

Unsatisfactory though the telephony and telegraph position is, that of television is far worse, since ionospheric transmission cannot be employed at all. The service area of a television station, even on a very good site, is only about a hundred miles across. To cover a small country such as Great Britain would require a network of transmitters, connected by coaxial lines, waveguides or VHF relay links. A recent theoretical study[1] has shown that such a system would require repeaters at intervals of fifty miles or less. A system of this kind could provide television coverage, at a very considerable cost, over the whole of a small country. It would be out of the question to provide a large continent with such a service, and only the main centres of population could be included in the network.

The problem is equally serious when an attempt is made to link television services in different parts of the globe. A relay chain several thousand miles long would cost millions, and transoceanic services would still be impossible. Similar considerations apply to the provision of wide-band frequency modulation and other services, such as high-speed facsimile which are by their nature restricted to the ultra-high-frequencies.

Many may consider the solution proposed in this discussion too far-fetched to be taken very seriously. Such an attitude is unreasonable, as everything envisaged here is a logical extension of developments in the last ten years—in particular the perfection of the long-range rocket of which V2 was the prototype. While this article was being written, it was announced that the Germans were considering a similar project, which they believed possible within fifty to a hundred years.

(*Reprinted from* Jet Propulsion, *April, 1955*)

Copyright, 1955, by the American Rocket Society, Inc.

Orbital Radio Relays

J. R. PIERCE[1]

Bell Telephone Laboratories, Murray Hill, N. J.

While orbital radio relays probably could not compete with microwave radio relay for communication over land, they might be useful in transoceanic communication. Three sorts of repeaters appear to be consistent with microwave art: (a) 100-ft reflecting spheres at an altitude of around 2200 miles; (b) a 100-ft oriented plane mirror in a 24-hr orbit, at an altitude of 22,000 miles; (c) an active repeater in a 24-hr orbit. Cases (a) and (b) require at 10-cm wavelength 250-ft-diam antennas and 100-kw and 50-kw power, respectively; in case (c), using 250-ft antennas on the ground and 10-ft antennas on the repeater, only 100 watts on the ground and 0.03 watt on the repeater would be required; in this case one should probably use smaller antennas on the ground. In cases (b) and (c) the problem of maintaining the correct orientation and position of the repeater is critical; perturbations by the moon and sun might cause the satellite to rock or wander prohibitively.

Figure 15.1. The first satellite proposals: A. C. Clarke (1945) and J. R. Pierce (1955).

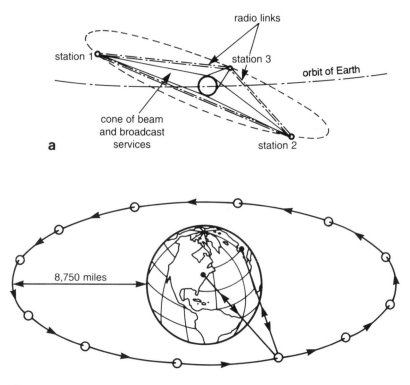

a

b

Figure 15.2. Synchronous and subsynchronous satellite orbits. (a) A. C. Clarke's (1945) satellite system proposal (3 satellites in 24-hour synchronous orbit at a height of 22,300 miles; transmission delay 0.3 seconds). (b) Proposal made at the IEE International Conference on Satellite Communication, London, 1962 (12 satellites in 8-hour subsynchronous orbit at a height of 8750 miles; transmission delay 0.13 seconds).

Clarke proposed three satellite "stations" in a circular equatorial orbit at a height of 22,300 miles above the Earth's surface and moving in a west-to-east path in space. At this height the orbital period is 24 hours and the satellites appear stationary when viewed from a west-to-east rotating Earth. Such an arrangement would provide virtually complete coverage of the globe, except for limited areas near the Poles. He envisaged the use of microwaves, noting their freedom from absorption in the atmosphere and ionosphere, with parabolic reflectors to concentrate the transmitted and received energy into narrow beams. Such an arrangement not only greatly reduces the satellite transmitter power required, it also by virtue of the "stationary" orbit simplifies the problem of pointing the highly directional receiving aerials at the satellite. He foresaw not only the use of satellites for point-to-point telecommunication services such as telephone and telegraph, but also the potential for direct broadcasting to homes.

His 1945 article in *Wireless World* refers to the possibility of both manned space stations and unmanned satellites "left to broadcast back to Earth." He discusses the use of energy radiated by the sun to power space stations and satellites via "solar engines" using low-pressure steam, and goes on to say that "thermo-electric and photo-electric developments may make it possible to use the solar energy more directly."

Clarke's proposals, made so many years before a satellite orbiting the Earth had become a reality, have been acclaimed as a work of genius for their simplicity, comprehensiveness, and remarkable resemblance to the satellite communication systems in use today. But, as Clarke has said, "the publication of the *Wireless World* article was received with monumental indifference"; he turned his attention and talents to other fields, notably to science fiction writing.

Meanwhile, on the other side of the Atlantic, J. R. Pierce, a scientist at the Bell Telephone Laboratories—remarkably also a science fiction writer of considerable repute*—had been discussing the possibilities of satellites for communication, but without knowledge of Clarke's 1945 proposals.

Dr. J. R. Pierce came to Bell Telephone Laboratories after receiving a Ph.D. from California Institute of Technology in 1936, and was Director of Electronic Research at Bell in the 1950s. He was at the forefront of almost every field of research in the telecommunication sciences, from acoustics and information theory to antennas and traveling-wave tubes.

He first put forward his ideas on communication satellites in a talk to the Princeton Section of the American Institute of Radio Engineers in 1954, later published under the title "Orbital Radio Relays" in the

---

*An article by Dr. J. R. Pierce, written under the pseudonym J. R. Coupling, and published in *Astounding Science Fiction* in March, 1952, mentions the possibility of interplanetary radio communication and reflections of radio waves from the moon.

periodical *Jet Propulsion* in 1955.[2] His paper discusses both passive satellites using spheres, dipoles, or plane mirrors as reflectors of radio waves, and active satellites with powered repeaters, in various orbits up to the synchronous orbit at 22,300 miles above the Earth's surface. He also refers to the problems of attitude control and station-keeping that would need to be solved, especially for satellites with directional aerials in the synchronous orbit. Pierce's views in 1954 on the utility of satellite communications were remarkably modest: "while orbital radio-relays probably could not compete with micro-wave radio-relays over land, they might be useful in trans-oceanic communication."

Pierce's realistic approach to satellite communication—which advocated first steps via nonsynchronous lower-orbit satellites, typified by Bell System's successful TELSTAR project of 1961—was in sharp contrast to the U.S. Army's project ADVENT which aimed in one step at a synchronous-orbit, station-keeping system with all of the problems of remote control that this entailed, and which was abandoned in 1962. With Rudolf Kompfner of Bell Labs (see Chapter 11) he drew attention in particular to the vital need for long-life active components, e.g., traveling-wave tubes, for use in satellites if viable commercial operation was to be achieved.

Pierce's book *The Beginnings of Satellite Communication,*[3] published in 1968, gives a lively account of the early history, and contains reproductions of both Clarke's 1945 paper and his own "Orbital Radio Relays" paper of 1955.

THE RUSSIANS ARE FIRST IN ORBIT, WITH SPUTNIK 1

There is no doubt that the launch of the grapefruit-size satellite Sputnik 1 from Tyuratam in the USSR on October 4, 1957, was a key event that sparked intensive efforts in the United States to launch satellites, notably but not initially successfully, by the military, who saw important possibilities for surveillance as well as communication. Because of the relatively low orbital height, signals from the low-power VHF transmitter on Sputnik 1 were readily receivable by radio amateurs and others throughout the world; this gave convincing proof of the satellite's existence and enabled its orbit to be readily followed.

The successful launch into orbit of Sputnik 1 was a convincing tribute to Russian prowess in rocketry and especially the multistage rocket concept attributed to Konstantin Tsiolkowsky (1857–1953), a Russian schoolmaster. By firing the stages in succession as increasing height was reached, and jettisoning each of the initial stages as its fuel was expended, greater orbital heights and payloads could be achieved for a given overall expenditure of rocket fuel and launch costs.

A SATELLITE FOR ALL TO SEE: THE ECHO BALLOON SATELLITE

The next important step was the launching, into an orbit 1000 miles high, of a 100-ft-diameter lightweight balloon satellite ECHO 1 by the U.S. National Aeronautics and Space Agency. NASA's original intention had been to use the balloon to measure the density of the upper atmosphere by observing the decay of the orbit—information of great importance in the design of future satellite systems.

J. R. Pierce and R. W. Kompfner of the Bell Laboratories realized that the balloon satellite with its metallized surface presented a valuable opportunity to explore the reflection of microwaves by a spherical passive reflector and to assess its value for communication puposes. To this end a joint project was established between NASA, the Jet Propulsion Laboratory (JPL; then a part of NASA), and the Bell Laboratories.[4]

ECHO 1 was successfully launched into a circular earth orbit on August 12, 1960, by a Delta rocket, the 1000-mile-high orbit giving an orbital period of about 2 hours. Microwave transmissions at 960 and 2390 MHz were made between the JPL Labs at Goldstone, California, and Bell Labs at Holmdel, New Jersey, using wide-deviation FM to transmit speech signals.

Since the attenuation over the path from the transmitting ground station via the scattered reflection from the satellite to the ground receiving station is very large, low-noise receiving equipment was essential. In the ECHO experiments this was achieved by using a "maser" liquid-helium-cooled preamplifier (*maser* is an acronym for "microwave amplification by stimulated emission of radiation"). When operated from a steerable horn-reflector aerial of the type originated by Friis and Bruce (see Chapter 11), the maser receiving system was sufficiently low-noise, about 19 degrees K, to detect the 3 degree K background noise radiation associated with the "big bang" and the creation of the universe—an observation for which A. A. Penzias and R. W. Wilson were awarded a Nobel Prize in 1978.

Also of interest in the experiments was the use of a "frequency following" FM receiver, invented by J. G. Chaffee of the Bell Labs more than 20 years earlier, which lowered the noise threshold of reception by several decibels.

In December, 1960, on the first pass of mutual visibility between Goldstone and Holmdel, a voice message from President Eisenhower was successfully transmitted. However, it was clear that passive reflection satellites, with their limited communication capacity—not much more than a single voice channel—had little commercial future.

The real value of the ECHO experiment was twofold. First, it was a valuable precursor to the development of the ground station equipment, rocket launching and satellite tracking techniques needed for commercially viable communication satellite systems using active satellites. Second, it showed the public that satellites were "for real"—no one who watched will forget the sight of the ECHO balloon, glowing in the light of sunrise and sunset, and moving majestically across the sky from horizon to horizon.

THE TELSTAR AND RELAY SATELLITE PROJECTS

The TELSTAR project was created and funded by AT&T and the Bell Telephone Laboratories as a private venture aimed at maintaining their preeminence in long-distance telecommunications. The RELAY project on the other hand was initiated by NASA, with the Radio Corporation of America as the prime contractor. NASA provided rocket launching facilities at Cape Canaveral for both projects and satellite tracking facilities at its Goddard Space Flight Center.[4]

The primary aims of both projects were the same, i.e., to demonstrate that active communication satellites could provide reliable, good-quality long-distance telephone, telegraph, data, facsimile, and television circuits—with the prospect of competing in quantity of circuits and cost with established long-distance transmission systems such as land microwave radio-relay and submarine coaxial cable systems. A further aim was to collect more scientific data on the space environment as an aid to design of future satellite systems.

The TELSTAR Project

The TELSTAR project was one of the major scientific and engineering achievements of the 20th century, establishing as it did the practicability of worldwide communication by satellites. While the overall planning, design, and management of the project were the responsibility of AT&T/Bell Labs, the British Post Office and the French PTT Administration agreed to provide ground stations in their respective countries and to carry out tests and demonstrations in cooperation with BTL.

The BTL project managers directly involved were A. C. Dickieson, Head of Transmission Systems Development, and E. F. O'Neill, TELSTAR Project Engineer.

TELSTAR was an active satellite launched in an elliptical orbit varying from 590 to 3500 miles in height, inclined at an angle of about 45° to the equator. The inclined orbit enabled a wide sampling of the space environment to be made, e.g., to assess the liability of satellite electronic components such as transistors to damage by the Van Allen radiation belts that surround the Earth, and the possibility of damage from micrometeorites. The orbital period was about 2.5 hours, the time of mutual visibility for ground stations on opposite sides of the Atlantic being some 30 minutes or less. The ground station aerials were thus required to track the moving satellite as it appeared above the horizon until mutual visibility ceased.

The near-spherical, 34.5-inch-diameter, 170-pound TELSTAR satellite was spin-stabilized, its rotation around a central axis giving it a gyroscopic stability so that its axis tended to remain in a fixed direction in space—thus minimizing the risk of loss of signal because of nulls in the directional patterns of the satellite aerials. Radiotelemetry was provided for satellite control purposes and to enable the transmission of data from the satellite. The electrical power to operate the communications, telemetry, and control equipment was provided by solar cells that converted sunlight into electrical energy stored in nickel cadmium cells both for normal operation and to maintain continuity when the satellite was in the Earth's shadow.

Figure 15.3. (Top) AT&T/Bell Telephone Laboratories TELSTAR satellite (170 pounds, 34.5-inch diameter) and (bottom) AT&T/BTL satellite ground station at Andover, Maine (showing 170-ft-long steerable horn aerial and 200-ft-diameter inflated radome).

The communication repeater on the satellite provided a bandwidth of 50 MHz, accommodating a frequency-modulated carrier bearing some 600 FDM telephone channels, or a television signal. It used a 2-watt-output traveling-wave tube and in all the satellite contained about 2500 solid-state devices, mainly transistors and diodes. The design is a tribute to Leroy C. Tillotson, a Bell Labs engineer whose internal BTL memorandum of August, 1959, proposed most of the characteristics eventually adopted for the TELSTAR satellite.[3]

The radio frequencies to be used were agreed on in principle by BTL with the cooperating ground station authorities (U.K. British Post Office and French PTT) on the basis of shared use with the frequency bands already recognized internationally for terrestrial microwave radio-relay systems, i.e., around 4 and 6 GHz. This understanding undoubtedly facilitated the agreement reached later by CCIR Study Group V (Space Systems) and adopted by the Administrative Radio Conference on frequency allocations for operational satellite communication systems, and the conditions to be observed to avoid mutual interference in the shared bands. (Frequencies used for TELSTAR were 4170 MHz for the down path and 6390 MHz for the up path, the lower frequency being used for reception at the ground station because of the slightly lower level of radio noise from the atmosphere.)

The ground station aerial designs differed markedly: for example, BTL used a large steerable horn-reflector under a 200-ft-diameter inflated radome at its ground station at Andover, Maine, and the British Post Office an 85-ft-diameter open parabolic dish-reflector at its Goonhilly, Cornwall, ground station. The French PTT at Plumeur Bodou, Brittany, used a replica of the BTL, Andover design.

BRITISH POST OFFICE CONTRIBUTIONS TO THE EARLY HISTORY OF SATELLITE COMMUNICATIONS

BPO responsibilities for overseas telephone communication with the United Kingdom, based initially on the use of shortwave radio systems and subsequently on submarine cables, gave it a strong interest in the satellite communications developments that were taking place on the opposite side of the Atlantic. The BPO Engineering Department, notably in its Radio Experimental Branch and Research Department at

Dollis Hill, NW London, had considerable experience in the design and use of microwave radio-relay systems (Chapter 11) that provided an excellent base from which to initiate work on satellite systems. The latter had the firm support of the then Engineer-in-Chief, Sir Albert Mumford, who established a team for this purpose led by Capt. C. F. Booth, Assistant Engineer-in-Chief, W. J. Bray, Head HQ Space Communication Systems Branch,* and F. J. D. Taylor, Head Radio Experimental Branch.

Much consideration was being given in the 1960s to the desirable orbits for a worldwide communication satellite system, taking into account the sometimes conflicting technical, operational, and economic factors involved. Early proposals included an initial system of 50 to 100 satellites in random orbit at a height of about 1000 miles, put forward by J. R. Pierce of the Bell Labs, influenced no doubt by the difficulties he then foresaw in achieving reliable station-keeping and attitude control in a synchronous, i.e., stationary, orbit system at a height of 22,300 miles.

Another problem anticipated with the synchronous orbit was the transmission delay of 0.3 seconds which accentuated the effect on talkers of echoes produced at the impedance mismatch frequently encountered at the junction between two- and four-wire local telephone distribution networks. This made it difficult for a would-be talker to "break in" on one already talking and led the author to propose a "subsynchronous" equatorial orbit with 12 satellites at a height of 8750 miles, where the orbital period would be exactly 8 hours, and the transmission delay reduced to 0.13 seconds. Such a system would of course entail tracking the moving satellites, but in a regular and simple pattern.[6]

Eventually the clear advantages of the synchronous orbit and "stationary" satellites prevailed over the "break-in" disadvantage, once the problems of accurate station-keeping and attitude control had been solved.

Realizing the importance and potential of satellite communications, the U.K. Government Cabinet Office arranged for a joint civil/military team to visit the United States in 1960 to investigate firsthand satellite

---

*Chapter 14 of the author's book *Memoirs of a Telecommunications Engineer*[5] contains an account of this exciting period from the viewpoint of one of the participants.

system developments. The outcome was an agreement with NASA, signed in February, 1961, to conduct satellite communication tests across the North Atlantic and an understanding that the BPO would build, at its own expense, a ground station for the purpose.

At about this time the BPO was engaged in technical studies with U.K. military establishments (the Royal Aircraft Establishment, the Army Signals Research and Development Establishment, and the Navy Surface Weapons Establishment) to examine the possibilities of a joint civil/military satellite communication system and to consider the possible use of the British "Blue Streak" rocket with German and French second and third stages, under the aegis of the European Launcher Development Organization, as a means for launching satellites in such a system. Although the outcome was nugatory in that the objectives of civil and military systems proved to be considerably different, and the Blue Streak rocket project had to be abandoned for a variety of economic, political, and technical reasons, the studies themselves proved very instructive for the BPO engineers in the fields of rockets, orbits, and payloads.[7]

THE BUILDING OF THE FIRST U.K. SATELLITE COMMUNICATION GROUND STATION AT GOONHILLY, CORNWALL (1961–1962)

The building of the first U.K. satellite communication ground station in one year, from a virgin site in July, 1961, to readiness for the TELSTAR tests in July, 1962, was perhaps an unexpected achievement for a Government civil service department such as the British Post Office. However, the BPO Engineering Department had a long history of handling large-scale radio and civil engineering projects, for example the high-power Rugby Radio Station, which gave it the experience and confidence to "do its own thing."

The building of Goonhilly involved much new and untried microwave technology such as low-noise maser and high-power traveling-wave tube amplifiers, and the construction of an 85-ft-diameter parabolic reflector aerial weighing over 870 tons, capable of being steered with great precision to track a rapidly moving satellite. It also involved design decisions and the coordination of many contributions from various sec-

Figure 15.4. Post Office Satellite Communication Ground Station Aerial No. 1 at Goonhilly, Cornwall (1962).

tors of industry and other Government departments, against a time scale that left no room for errors of judgment. The story of this achievement is described in Ref. 5; the design of the station and the results obtained are described in Refs. 8 and 9.

The choice in 1960 of the ground station site at Goonhilly on the Lizard Peninsula, Cornwall, was based on the desire to facilitate transatlantic communication—which indicated a location as far west as was practicable—and the need to avoid interference to and from terrestrial microwave radio-relay systems using the shared 4- and 6-GHz frequency bands. Almost by accident the chosen site was near Poldhu from which Marconi made the first long-wave radio transmissions to Newfoundland in 1901 (Chapter 7).

The civil engineering design and construction of the 85-ft-diameter parabolic reflector aerial at Goonhilly was due to Dr. Tom Husband, whose firm Husband and Co. had been responsible for the 240-ft-diameter radiotelescope used for radioastronomy at Jodrell Bank, Cheshire. The construction had to be massive to withstand the gales experienced at times on the windswept Lizard Peninsula, and yet had to be capable of precision steering, achieved by driving it in azimuth by a substantial Reynolds chain—derided by one newspaper at the time as a "bicycle chain"! The choice of an open-dish design of aerial, unprotected by a radome as was used by the Bell Labs at Andover and Plumeur Bodou, proved to be advantageous both in terms of cost and operationally, as the later tests showed.

THE ATLANTIC IS BRIDGED BY SATELLITE, JULY, 1962

The TELSTAR satellite was launched by NASA from Cape Canaveral at 0835 GMT on July 10, 1962. There followed between the 10th and 27th of July an intensive period of tests and demonstrations that included the transmission of monochrome and color television signals, multichannel telephony, data, and facsimile signals.

On the first pass of TELSTAR on July 10, the Goonhilly ground station experienced a difficulty related to the reversal of the polarization of the aerial, which did not match that of the downcoming wave from the satellite and which arose from an ambiguity in the definition of "right-handed circular polarization." Thus, the honor of the first direct reception of television signals from the United States fell to the French PTT at Plumeur Bodou; however, the polarization error at Goonhilly was soon corrected and on July 11 the first television transmissions to the United States were made from Goonhilly; they were reported by Andover as of "excellent quality" and rebroadcast throughout the United States.

These exciting and dramatic television exchanges were followed on July 12 by the first transatlantic telephone conversations by satellite between BPO engineers at Goonhilly and AT&T/BTL engineers at Andover. On July 14, Sir Ronald German, Director General of the British Post Office, talked via TELSTAR with Mr. McNeely, President of AT&T, who spoke of "the thrill of communicating through a dot in space." Queen Elizabeth referred to TELSTAR as "the invisible focus of a million eyes."

The early television transmissions across the Atlantic had been in monochrome and some doubts existed regarding the possible effect of Doppler frequency shift caused by satellite movement on the chrominance subcarrier of a color television signal. However, with the help of color picture generating equipment supplied by the BBC, the first color television transmissions were made by BPO engineers from Goonhilly on July 16/17, 1962, using 525-line NTSC standards. The quality of reception at Andover was rated "very good" and gave further confidence in the transmission quality of the satellite link.

The author summed up the results of the tests as follows:

> The results obtained from the TELSTAR demonstrations and tests to date (end 1962) have confirmed the expectation that active communication satellites could provide high-quality stable circuits both for television and multi-channel telephony. The very good results obtained with colour television signals—a stringent test of any transmission system—and the tests with 600 simulated telephone circuits, are particularly noteworthy.[9]

THE SYNCHRONOUS ORBIT IS ACHIEVED: SYNCOM

Credit for achieving the first ever synchronous orbit of a satellite goes to the engineers of the Hughes Aircraft Company: in July, 1963, they succeeded in placing a satellite in the synchronous orbit at a height of 22,300 miles—thus fulfilling the first stage of A. C. Clarke's vision of 1945.

The Hughes proposals were sponsored by NASA as Project SYNCOM; they involved a satellite of the minimum practicable weight—only 78 pounds—because of the restrictions imposed by the Thor–Delta rocket launcher on the payload that could be put into the synchronous orbit. The need to keep the weight low meant that storage batteries to maintain operation in the Earth's shadow, when the solar cells failed to receive sunlight, could not be included. The SYNCOM satellite contained an apogee (jet) motor for injecting it into the final synchronous orbit after separation of the satellite from the rocket launcher. It was spin-stabilized, and used gas jets on the satellite controlled by radiotelemetry from the ground to adjust its position and attitude in orbit.

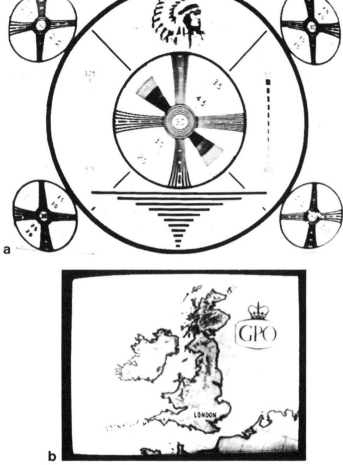

Figure 15.5. The first tests via the TELSTAR satellite in July, 1962. (a) "Indian Head" test card received at Goonhilly from AT&T/BTL satellite ground station, Andover, Maine, via TELSTAR on July 11, 1962. (b) Map of the United Kingdom transmitted from Goonhilly via TELSTAR on July 11, 1962.

As viewed from the ground, even satellites in the synchronous orbit are not completely "stationary"; they generally describe a narrow "figure eight" pattern above and below a mean position near the equator, and when high-gain, narrow-beam ground station aerials are used they have to be tracked, albeit over a limited angular range.

The radio equipment on SYNCOM comprised a single repeater, receiving signals from the ground at 7400 MHz and transmitting at 1800 MHz with a power of 4 watts from a traveling-wave tube amplifier. Signal-to-noise ratio considerations meant that the repeater bandwidth had to be restricted to about 0.5 MHz; nevertheless, multichannel telephony, facsimile, and data transmission were successfully demonstrated.

The success of Project SYNCOM paved the way for the next generation of satellites by demonstrating unequivocally that the synchronous orbit was achievable, and that station-keeping and attitude control were practicable.

THE ONWARD MARCH OF SATELLITE COMMUNICATIONS

From the successes of TELSTAR and SYNCOM the satellite story moves out of the hands of its scientific and engineering innovators and creators: powerful political and financial forces were at work, both nationally and internationally, and the story becomes increasingly complex.

In August, 1962, the U.S. Government, realizing the immense significance of these developments in satellite communications, passed the Communications Satellite Act, which created a Communications Satellite Corporation (COMSAT) with the aim of securing American control of the manufacture, launching, and marketing of satellites for international telecommunications.

It became clear that earlier procedures for the establishment of international telecommunication links such as submarine cables, which rested on bilateral agreements between the participating PTT authorities in the countries concerned with ratification by Governments, now had to be replaced by multilateral agreements which necessarily involved COMSAT. This was recognized by the 22-nation European Conference of Postal and Telecommunication Administrations (CEPT) in December,

1962, which set up in 1964 an Interim Communications Satellite Committee in which the United States had through COMSAT a majority ownership share.

The first successful commercial communication satellite launched in April, 1965, by COMSAT was the synchronous orbit INTELSAT ll (Early Bird), which provided 240 telephone circuits or one television channel between Europe and the United States. From this modest beginning, communication satellites were, within two decades, providing the bulk of the world's intercontinental and large numbers of transcontinental telephone and television links. Satellite capacities came to be measured in tens of thousands of telephone circuits and tens of television channels. Specialized satellites for maritime ship-to-ship and ship-to-shore communication, and communication to and from aircraft were launched. And television broadcasting directly from high-power satellites to viewers' homes came into being.

And, far beyond even A. C. Clarke's 1945 prophetic forecast, spacecraft in the 1980s transmitted, using image storage and slow-scan techniques, high-quality television pictures of the moon, Saturn's rings, and the moons of the most distant planets of the solar system.

It must have been immensely gratifying for the innovative scientists and engineers involved in the early stages of this remarkable development of the telecommunicators art to have seen their ideas and work come to such fruition within their lifetime.

REFERENCES

1. A. C. Clarke, "Extra-Terrestrial Relays: Can Rocket Stations Give World-wide Radio Coverage?" *Wireless World*, October 1945.
2. J. R. Pierce, "Orbital Radio Relays," *Jet Propulsion*, April 1955.
3. J. R. Pierce, *The Beginnings of Satellite Communication* (San Francisco Press, 1968).
4. *A History of Engineering and Science in the Bell System: Transmission Technology (1925–1975)* edited by E. F. O'Neill (Bell Laboratories, 1985), Chapter 13.
5. John Bray, *Memoirs of a Telecommunications Engineer*, Chapter 14, 1982. Available from British Telecom Showcase, Upper Thames Street, London.
6. W. J. Bray, "Satellite Communication Systems," *Post Off. Electr. Eng. J.*, 55, July 1962.

7. "A Study of Satellite Systems for Civil Communication," Royal Aircraft Establishment Report No. GW26, March 1961. Published in shortened form in *Telecommunication Satellites*, edited by K. W. Gatland (Iliffe Books Ltd., 1964).
8. F. J. D. Taylor, "Equipment and Testing Facilities at the Experimental Satellite Ground Station, Goonhilly Downs, Cornwall," *Post Off. Electr. Eng. J., 55,* July 1962.
9. W. J. Bray and F. J. D. Taylor, "Preliminary Results of the Project Telstar Communication Satellite Demonstrations and Tests," *Post Off. Electr. Eng. J., 55,* October 1962.

Pioneers of Long-Distance Waveguide Systems

An Unfulfilled Vision

GUIDED ELECTROMAGNETIC WAVES: THE TELECOMMUNICATION HIGHWAYS OF THE FUTURE

The possibility that electromagnetic (EM) waves, whether as microwaves or light waves, might be propagated over long distances on dielectric rods or in metal tubes with or without a dielectric filling was studied from motives of scientific and mathematical interest, long before applications for long-distance communication seemed either practicable or useful. As long ago as 1897, Lord Rayleigh in the United Kingdom had shown, from certain solutions of Maxwell's field equations (Chapter 2), that EM waves could propagate freely inside hollow metal tubes provided that the wavelength was appreciably shorter than a cross-sectional dimension of the rod or tube.[1] This meant in effect that there was a "cutoff" frequency below which EM waves would not propagate;

for rods or tubes of a convenient size, e.g., a few centimeters across, frequencies of many gigahertz were required for free propagation. In the absence of suitable sources and detectors for such frequencies, it was not until the 1930s that pioneering experimental studies of centimeter-wavelength guided waves began to be made in the United States by George C. Southworth of AT&T/Bell Labs.[2] (EM waves on wire pairs or coaxials are also "guided," but without a lower limit of frequency.)

Clerk Maxwell had demonstrated mathematically in the 1870s that light consists of EM waves, differing from microwaves only in that the wavelengths involved were extremely short, 0.4 to 0.75 μm (1 μm is one ten-thousandth of a centimeter) in the case of visible light, and the frequencies correspondingly high. The term "light" is conventionally extended to include EM waves in the infrared region of the spectrum, from 0.75-μm wavelength to the millimeter-wavelength microwaves, and also the ultraviolet region below 0.4 μm. Having established the identity of light with EM waves, it became clear that in principle light waves too could be guided in glass fibers of diameter greater than a wavelength. And thus there was at least a possibility that "optical fibers" transmitting visible or infrared light could be used for communications. However, this possibility was not investigated experimentally until the 1960s when K. C. Kao and G. A. Hockham of Standard Telecommunication Laboratories, England, published their first results.[4]

The requirements of intercity links for multichannel telephony and television transmission in the period up to the 1980s had been met in Europe, the United States, and most other countries by coaxial cable and microwave radio-relay systems. However, the exponential growth of telecommunication traffic foreseen beyond that period made it worthwhile to study intensively guided-wave systems, with their potentially high communication capacity on a single tube or glass fiber. The earliest approaches involved centimeter-wavelength guided-wave systems using copper tubes, but as the possibility of achieving low-loss transmission of light in glass fibers began to be realized, the pendulum began to swing in favor of optical systems, with all of the operational advantages of a mechanically convenient and compact medium they offered.

THE WAVEGUIDE PIONEERS

The accolade for farsighted vision of the possibilities of waveguides for long-distance communication and a determined attack—in the face of considerable practical and managerial difficulties—on the design problems involved, must go to George C. Southworth of AT&T/Bell Telephone Laboratories.

George C. Southworth

The story of the early waveguide work is well told in Southworth's book *Forty Years of Radio Research*,[2] in which Lloyd Espenschied of BTL has written in the Foreword:

> Related are not only the subtle technical problems and how they were met in long-sustained effort, but also the human side of the picture, the motivations, trials, and tribulations in a great stratified organization, periodic victories, and frustrations.

Southworth's first contact with guided wave phenomena came when, as a student at Yale University in the 1920s, he was experimenting with high-frequency waves propagated in a water-filled rectangular trough and observed unexpected wave patterns. The high dielectric constant of water (c. 80) had increased the cutoff wavelength to a point where a variety of wave patterns could exist. These he recognized as similar to the types of wave predicted mathematically by O. Schriever at the University of Kiel in 1920, in a study of the propagation of EM waves on dielectric rods.[5] (As noted above, Lord Rayleigh had much earlier, in 1897, made similar predictions.)

Following a strong instinct to pursue the matter further, and with little initial managerial support, Southworth, then a research engineer with AT&T, began work on dielectric-filled metal pipes as waveguides in 1931. His notebook of November 10, 1931, contains an entry proposing their use as a transmission medium. The following summary of his subsequent work is based on Ref. 2 and the account in the BTL *History of Engineering and Science*.[3]

By March, 1932, Southworth had set up at Netcong, New Jersey, water-filled waveguides 6 and 10 inches in diameter, and a few feet long, energized from an oscillator covering 150 to 250 MHz and equipped with launching and detecting devices. With this equipment he was able to identify the various EM modes and measure wave phase and other characteristics. Remarkably, this work was at first discouraged by AT&T mathematician J. R. Carson on the grounds that "it was not practical"—perhaps because the Laboratories were then struggling with the early coaxial cables and were very much aware that the cable attenuation

Figure 16.1. Waveguide pioneer G. C. Southworth (right) and waveguide field patterns (opposite).

increased with frequency to a degree which at the time limited the highest usable frequencies to a few megahertz.

However, Southworth's experimental results and his determination to take the work further stimulated a more detailed mathematical analysis of EM wave propagation in hollow metal pipes. This was carried out at AT&T by S. A. Schelkunoff and Sally P. Mead.

S. A. Schelkunoff at Bell Laboratories had earlier studied analytically the propagation characteristics of coaxial cables and also turned his attention to guided waves in metal pipes. Mead and Schelkunoff's studies revealed in May, 1933, a remarkable property of one particular EM wave mode—now known as the circular-electric TE01 mode—in which the attenuation in a truly circular guide decreased inversely as the 3/2 power of the frequency.[6] In practice, surface and other imperfections cause the attenuation eventually to increase with frequency, but only after a very wide useful frequency band has been achieved.

It appears that credit for the first formulation of this remarkable result should go to Sally P. Mead, while S. A. Schelkunoff seems to have been the first to realize the significance of the asymptotic behavior with frequency.

By 1935 a 6-inch copper pipe waveguide 1250 ft long had been built at the Bell Laboratories, Holmdel, and with it Southworth demonstrated

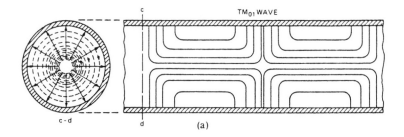

(a)

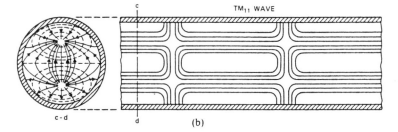

(b)

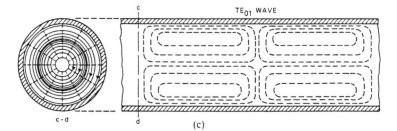

(c)

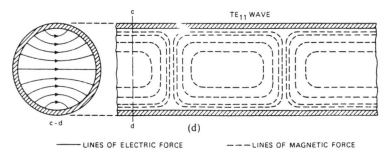

(d)

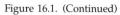

Figure 16.1. (Continued)

transmission of the TE11 mode wave. However, at that time, the TE01 low-loss mode was above the frequency range of available test equipment and it was not until the late 1930s that the attenuation of this mode could be measured. The measured values were at first considerably above the theoretically predicted values, because of waveguide imperfections that resulted in the conversion of the low-loss TE01 mode to lossier modes.

It took many years of painstaking research and development before waveguides with losses approaching the theoretically achievable losses with the TE01 mode were achieved. Perhaps the more immediate value of the guided-wave work of the 1930s was the impetus it gave to the microwave radar developments in the war years, and to microwave radio-relay systems in the postwar years.

MILLIMETER WAVEGUIDE DEVELOPMENT IN THE BELL SYSTEM

Work on TE01-mode guided-wave transmission was resumed at Bell Laboratories after the end of the war, with the expectation that it would make available a long-distance transmission medium of far wider bandwidth than was offered by coaxial cable or microwave radio systems. It was early realized that, for a practicable system, the waveguide diameter would need to be no more than a few centimeters; it followed that wavelengths in the millimeter range would be required, i.e., frequencies of more than 30 GHz. It was also envisaged that, although other types of modulation were possible, the wide bandwidth available would enable high-bit-rate digital modulation to be used with all of the other advantages that this conferred.[8]

During the 1950s the problems of mode conversion from the TE01 mode to unwanted modes, resulting from waveguide imperfections and bends, were intensively studied. Mode conversion not only increased waveguide losses, it also caused pulse waveform distortion as a result of the different propagation velocities of the unwanted modes compared with the low-loss TE01 mode. This pointed clearly to the advantages of PCM digital modulation, with its inherent resistance to such distortion.

Improvements in waveguide design were pursued, not only to reduce losses but also with the aim of relaxing the laying requirements with regard to straightness and curvature at bends. Bends were particu-

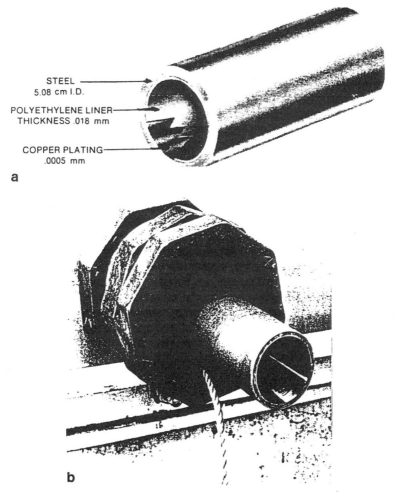

Figure 16.2. Waveguide designs for field trials. (a) Bell System dielectric-lined millimetric waveguide; (b) BPO/Telecom helix waveguide in duct.

larly troublesome in untreated copper pipe, because of conversion from the transverse-electric TE01 mode to the transverse-magnetic TM11 mode which had the same wavelength in the guide. One approach was to replace the copper tube with a continuous fine-wire helix which had little effect on the TE01 mode with its almost zero transverse-electric field at the circumference, but which absorbed other modes with a relatively strong circumferential electric field. Another approach was to line the copper tube with a dielectric which had a similar effect—and which was cheaper to manufacture. Both of these solutions had become available in the early 1960s, but by then the problems of repeater design loomed large.

Early experimental repeater designs used thermionic valves as "backward-wave" oscillators (due to R. Kompfner) and millimeter traveling-wave amplifiers, but the problems of large power requirements and close manufacturing tolerances indicated strongly the need for solid-state devices in a practicable system. The breakthrough came in 1963 when J. B. Gunn at IBM demonstrated microwave oscillations in gallium arsenide and indium phosphide diodes. Semiconductor devices developed by Bell Labs included the "IMPATT" (Impact Avalanche Transit Time Diode) as a solid-state millimeter-wave power source, and the "LSA" (Limited-Space Charge Accumulation) diode. These formed the basis of an all-solid-state regenerative repeater, i.e., one capable of accepting weak binary pulses at bit rates on the order of 300 Mbit/sec, reshaping and amplifying them for onward transmission.

By 1969 the development of waveguide, broadband channel multiplexing components and repeaters had reached a point where a field trial, with the objective of proving the design of a millimeter waveguide system suitable for the large volume of intercity telecommunications traffic predicted for the 1980s and beyond, could be conducted. The performance achieved by 1975 was as follows:

Voice channels per waveguide	238,000[a]
Broadband channel bit rate	274 Mbit/sec
Repeater spacing	31–37 miles
Waveguide diameter	6 cm

[a]In 124 broadband channels, 62 in each direction.

The field trial over a 14-km route used dielectric-lined waveguide with helical mode suppressors at intervals of about 0.5 miles. The overall loss of the waveguide was remarkably low over a frequency band from 38 GHz to 104.5 GHz, varying from 1 dB/km at band edges to 0.5 dB/km at the center.[9]

The massive and sustained effort on waveguide system research and development by AT&T and Bell Laboratories had demonstrated unequivocally that a millimeter waveguide system was practicable, could provide communication capacity far beyond existing coaxial cable and microwave radio-relay systems, and at lower unit cost. Telecommunication authorities in Great Britain, France, Germany, and Japan had followed the Bell Laboratories work with great interest and some had conducted similar development work; the British Post Office Research Department waveguide system field trial is described below. However, by the 1970s traffic growth in the United States had slackened and a competing system—optical fibers—was beginning to make its appearance. Perhaps the last word from the United States on waveguide systems should be left to the pioneer Southworth who in 1962 had written:

> Almost from the first, however, the possibility of obtaining low attenuations from circular electric waves, carrying with it the possibility of vastly wider bands of frequencies, appeared as a fabulous El Dorado always beckoning us onward.

MILLIMETER WAVEGUIDE DEVELOPMENT BY THE BRITISH POST OFFICE

Beginning in the mid-1960s scientists and engineers of the British Post Office, initially at the PO Research Department, Dollis Hill, NW London, and later at the PO/British Telecom Research Centre, Martlesham, Suffolk, carried out a comprehensive study of millimeter waveguides operating in the TE01 mode, leading to a field trial in 1974. In this the BPO was motivated by the same considerations as the Bell System, i.e., to cater in the most efficient way possible for the large and exponentially increasing volume of telephone, data, and visual telecommunication traffic between major cities foreseen for the 1980s and beyond.[10]

The work received considerable support from U.K. industry, notably from Standard Telecommunication Laboratories, Harlow, in the development of lasers and other solid-state high-bit-rate devices, and from British Insulated Callenders Cables who developed manufacturing facilities for the PO-designed lightweight waveguide. A group at University College, London, under the leadership of Professor E. M. Barlow, also contributed with theoretical and experimental studies of guided-wave propagation.

The BPO team on waveguide system research and development was led by Robert W. White and Bill Ritchie. R. W. White had an extensive earlier involvement in the development of radio-relay and satellite communication systems for the Post Office (Chapter 15). Bill Ritchie, who later received a Martlesham Medal for his work on broadband local distribution systems (Chapter 19), was responsible for the creation of the unique PO lightweight waveguide. For its field trial the PO opted for an all-helix type of guide in contrast to Bell Labs' mainly dielectric-lined waveguide with its rigid steel tube. The 5-cm-diameter fine-wire helix was supported by an epoxy-resin-impregnated glass-fiber tube, giving a precisely dimensioned, lightweight structure with great mechanical and electrical stability. A critical design factor was the spacing of an aluminum foil boundary around the wire helix, which much reduced mode-conversion losses and yielded a smoother overall loss/frequency characteristic. Also important was the low-loss bend design technique developed by Colin South based on critical dimensioning of the dielectric lining and which enabled losses in bends of only 1 meter radius to be reduced to less than 0.1 dB.

A mechanically simple and electrically efficient jointing technique was devised, which enabled sections of waveguide a few meters long to be readily transported and quickly connected under field conditions.

The 5-cm-diameter waveguide had an effective bandwidth extending from 30 to 110 GHz. Fully equipped with broadband channels operating at 100 to 500 Mbit/sec, it provided for up to 300,000 telephone circuits or 200 television channels, or a combination of these.

The field trial, which began in 1974, was over a 14-km route alongside a main trunk road in Suffolk with a variety of straight and curved sections. The lightweight guide was laid in a standard PO cable duct without special supports, its mechanical characteristics enabling it to flex smoothly and gradually. The overall attenuation of the 14-km length as laid was less than 3 dB/km over the 30 to 110 GHz band.

The trial demonstrated that a low-loss millimeter waveguide system was practicable, and that the performance objectives were achievable and stable. It showed that the waveguide could be manufactured in quantity, readily transported, and installed under field conditions without excessive constraints as to curvature and bends.

The decision to close down the field trial and further work on millimeter waveguides in 1976 was, in the light of the progress that had been made on optical fibers, a correct decision. Nevertheless, it was a sad moment for the PO/BT scientists and engineers who had labored so long and successfully to create a practicable and economic waveguide system—at that moment they must have felt a strong bond of sympathy with their friends in Bell Labs who had to accept a similar decision! But perhaps one useful outcome was the impetus the work gave to the development of high-bit-rate digital techniques and solid-state devices that later found application in optical fiber and millimeter radio-relay systems.

REFERENCES

1. Lord Rayleigh, "On the Passage of Electric Waves through Tubes, or the Vibrations of Dielectric Cylinders," *Philos. Mag.*, Ser. 5, *43*, February 1897.
2. G. C. Southworth, *Forty Years of Radio Research* (Gordon and Breach, New York, 1962).
3. *A History of Engineering and Science in the Bell System: Transmission Technology (1925–1975)*, edited by E. F. O'Neill (Bell Laboratories, 1985), Chapter 7.
4. K. C. Kao and G. A. Hockham, "Dielectric Fibre Surface Waveguides for Optical Frequencies," *Proc. IEE, 113*, No. 7, July 1966.
5. O. Schriever, "Elektromagnetische Wellen an Dielektrischen Drahten," *Ann. Phys.*, Ser. 4, *63*, December 1920.
6. J. R. Carson, S. P. Mead, and S. A. Schelkunoff, "Hyper-Frequency Waveguides—General Considerations and Theoretical Results," *Bell Syst. Tech. J., 15*, April 1936.
7. G. C. Southworth, "Hyper-Frequency Waveguides—General Considerations and Experimental Results," *Bell Syst. Tech. J., 15*, April 1936.
8. Ref. 3, Chapter 20.
9. "Millimeter Waveguide System," Special Issue *Bell Syst. Tech. J., 56*, December 1977.
10. John Bray, *Memoirs of a Telecommunications Engineer,* Chapter 15, 1982. Available from British Telecom Showcase, Upper Thames Street, London.

17

Pioneers of Optical Fiber Communication Systems

The First Transatlantic System

THE OBJECTIVE AND THE CHALLENGE

The objective facing the pioneers was the development of a transmission system offering greater communication capacity at lower cost than existing media such as coaxial cable and microwave radio-relay systems, in a mechanically flexible and compact form that could readily be used in the field.

Hair-thin glass fibers transmitting coherent light offered the possibility of almost unlimited communication capacity. Light with a wavelength on the order of 1 μm has a frequency of 300 million MHz: if only 1% of this frequency is effectively used, the corresponding communication bandwidth is 3 million MHz, wide enough to accommodate millions of telephone circuits or hundreds of television channels on a single fiber. Furthermore, the use of light as a carrier meant complete freedom from electromagnetic interference such as that originating from electrical machinery, power lines, and radio transmissions.

The technological challenges involved were substantial; first was the development of low-loss glass, starting from an initial level of 1000 dB/km or more, reducing it to a usable 10 dB/km and eventually to 1 dB/km or less. And this had to be achieved in a fiber structure in which the refractive index distribution between an inner core and an outer glass cladding effectively guided a light wave. Means had to be developed for continuously drawing glass fibers in lengths of tens of kilometers or more as a basis for a practicable and economic manufacturing process. The next challenge was the development of solid-state laser light sources and photodetectors, suitable for operation at pulse bit rates of 100 Mbit/ sec or more, and with the lives needed for commercial operation.

The possibilities of optical fibers began to be realized in the late 1960s, stimulated by the pioneering work of K. C. Kao at Standard Telecommunication Laboratories (STL), England. By the early 1970s the research laboratories of the British Post Office, Dollis Hill, and STL in England, the American giant Corning Glass, AT&T, and ITT in the United States, and NTT in Japan had begun intensive studies of the problems of creating a viable optical fiber system technology.

THE FIRST STEP

K. C. Kao's classic paper of 1966 presents the first comprehensive scientific, mathematical, and experimental study of the transmission of electromagnetic waves at optical frequencies by a circular section dielectric fiber with a refractive index higher than the surrounding medium; as such it provided not only a stimulus to further work but also a sound foundation for it.[1]

Dr. Charles K. Kao was born in Shanghai, China, in 1933. He received B.Sc. and Ph.D. degrees from London University and joined Standard Telecommunication Laboratories, England, where his work on optical fibers was carried out. His analysis of wave propagation follows that of Lord Rayleigh in 1897 in its use of Maxwell's equations (Chapter 2) to define a family of magnetic H and electric E modes, and a family of hybrid HE modes.

In particular, Kao identified and demonstrated experimentally the lowest-order hybrid HE11 mode with which it is possible to achieve single-mode operation by reducing the diameter of the guiding structure sufficiently. This was later to prove important in achieving high-bit-

Figure 17.1. Charles Kao, who, with G. A. Hockham, led the way to optical fiber systems (1966).

rate transmission, as compared with multimode operation which led to pulse waveform distortion and lower usable bit rates.

The paper goes on to discuss energy losses related to radiation and bending, scattering from imperfections, and absorption from impurities in the glass; it explores the variation of these losses with wavelength and reveals the existence of absorption bands related to molecular resonances in various types of material. In the case of fused quartz, for example, his results indicated an optimum wavelength of about 0.65 μm. (Later work has indicated even lower losses at about 1.3 μm.)

Kao's paper does not give measured losses for optical fibers; however, because of the imperfections of glasses available at the time, these were probably on the order of 1000 dB/km. He concludes:

(1) Theoretical and experimental studies indicate that a fibre of glassy material in a cladded structure with a core diameter of about λ (the light wavelength) and an overall diameter of 100 λ represents a possible prac-

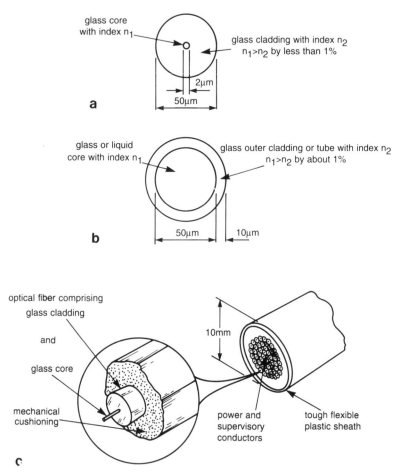

Figure 17.2. Single- and multimode optical fibers. (a) Monomode fiber; (b) multimode fiber; (c) multifiber cable. Fiber dimensions shown are for light wavelength of about 0.7 μm. For longer wavelengths, e.g., 1.3 μm, the dimensions would about double.

tical waveguide with important potential as a new form of communication medium.

(2) The refractive index of the core needs to be about 1% above that of the cladding.

(3) The realisation of a successful fibre waveguide depends, at present, on the availability of suitable low-loss dielectric material. The crucial mate-

rial problem appears to be one that is difficult, but not impossible, to solve. Certainly the required loss figure of around 20 db/km is much higher than the lower limit of loss imposed by fundamental mechanisms.

Time has shown that Kao's concepts of the optical waveguide and his expectations for it were not only sound but have been realized in full measure in operational systems throughout the world.

THE CORNING GLASS CONTRIBUTION

It was to be expected that American Corning Glass—the largest commercial organization in the world concerned with the development and manufacture of glass and glass products—would take an early interest in the possibilities of optical fibers. This interest began in 1966—significantly the date of publication of Kao's paper—when Corning scientist William Shaver visited London and discussed with engineers of the British Post Office the possibilities of optical fibers for telecommunications. The story of what followed from that beginning is well told in an extensive interview with Corning staff in 1989 by a consulting firm concerned with the evolution of strategy to protect U.S. global business interests, recorded in Ref. 2.

The task facing Corning was daunting indeed: the light loss in even good-quality glass was some 20 dB/m, compared with the objective of no more than 20 dB/km. It was a task that took more than 10 years to complete to the stage of a pilot manufacturing plant capable of making kilometer lengths of glass fiber of acceptable quality—an achievement that involved not only a long-sustained and determined effort in the face of many setbacks, but an original approach to fiber design that differed markedly from competing designs. Much of the credit for this must go to Robert Maurer, who was born in 1924 in Arkansas where he gained a B.S. degree in physics in 1948, followed by a Ph.D. at MIT in 1951. He joined Corning in 1952 as a physicist, later becoming Manager for Applied Physics Research and Special Projects, and Research Fellow. In addition to many technical publications on optical waveguides, he holds some 16 patents including the two key patents on those in current use throughout the world. His honors and awards include the Morris Liebmann Award of the U.S. Institution of Electrical and Electronic Engineers, the L. M. Ericsson International Prize for

Figure 17.3. Three scientists at Corning Glass, USA, who developed the first low-loss optical fiber. (Left to right: Dr. D. B. Keck, Dr. R. D. Maurer, and Dr. P. Schultz.)

Telecommunications awarded by the Swedish Academy of Engineering, and the American Physical Society International Prize for New Materials; he was elected to the U.S. Academy of Engineering in 1979.

His principal colleagues in the optical fiber work were Pete Schultz and Donald Keck. Schultz was a senior Corning scientist who specialized in creating new chemical formulas for glass, while Keck tackled the problems of measuring fiber purity and drawing glass into a fiber. With Maurer they made a determined and capable research team.

Following William Shaver's visit with the British Post Office engineers in 1966, he put the problem of creating a commercially viable optical fiber to William Armistead, Corning's Director of Research. Armistead realized the massive difficulties involved but saw the long-range potential of optical fibers and gave Maurer the go-ahead, realizing that there was mounting competition from other laboratories in England, West Germany, and Australia as well as the United States.

Corning had gained earlier experience from defense projects with fused silica, the purest glass then known. But it seemed at first most unpromising for an optical waveguide. First, the melting point was extremely high—some 2000°C, more than twice the melting point of steel—and this would make drawing into a fiber very difficult.

Second, the refractive index of fused silica was the lowest of known glasses. And as Kao had pointed out, the fine inner glass core of an optical waveguide has to have a refractive index slightly greater than that of the glass cladding, if light is to be effectively guided.

While Corning's competitors—including the British Post Office research team—were concentrating on refining the impurities out of optical glass, Maurer decided to stay with fused silica, a decision that later proved to have been critical for ultimate success.[3]

The key to the refractive index problem was chemically to dope the material of the inner core to raise its refractive index—as was remarked at the time, while the competition was refining chemicals out, Maurer was putting them in!

There remained the problem of drawing glass into a hair-thin fiber with the right dimensions of inner core and cladding. The well-tried solution used by most of the competition was to take a glass rod, surround it with cylindrical glass cladding of lower refractive index, heat the two in a vertical furnace, and draw from the lower melted tip—a process that maintained the relative dimensions of core and cladding but that was difficult to achieve with fused silica. The brilliant solution put forward by Schultz and Keck was to turn the process inside out—to start with a hollow tube of fused silica and apply the fused silica core material by vapor deposition, a technique that Corning had used successfully in other applications. In April, 1970, after 4 years of endeavor and many frustrating setbacks, a breakthrough was achieved and Donald Keck was able to demonstrate to Bill Armistead a meter length of laboratory-made fiber with a loss below the then target of 20 dB/km; Keck's laboratory notebook says:

Attenuation equals 16 db. Eureka!

As Keck said at the time, "he could just about feel the spirit of Edison in the laboratory."

But this, vital as it was, was only a beginning; techniques for manufacture of long lengths of fiber had to be developed and a market found among the telecommunication equipment manufacturers of the world.

It fell to Charles Lucy, an MIT graduate who joined Corning in 1952, to supervise the building of an optical fiber manufacturing plant capable of producing long lengths of fiber in a form suitable for use by telecommunication system designers. He had also to find a market for the product.

There was at first a marked reluctance to buying Corning fiber by overseas telecommunication equipment manufacturers and their cable suppliers, although a number of patent licensing agreements were made with firms in England, France, Italy, and Germany. Meanwhile Corning continued their research, achieving 4 dB/km in 1975, in fiber lengths up to some 4 km. Strenuous efforts were made to improve the manufacturing process, reduce costs, and increase the fiber length. By 1977 a new manufacturing process had been achieved in which a rod was sprayed with vaporized core glass and cladding sprayed on that; the rod was then removed and the fiber pulled from the double cylinder that was left. This process lent itself to the production of single-mode fiber with its ultrafine core—an essential for achieving the maximum communication capacity. Furthermore, the loss had been reduced to an incredible 0.3 dB/km. By 1988 the manufacturing plant was capable of producing up to 100-km lengths of single-mode low-loss fiber at a very competitive cost—and the door was open to the world market. However, the achievement of a majority share of that market involved much litigation against overseas and some U.S. firms for infringement, litigation in which Corning was notably successful through its ownership of optical fiber master patents.

The Corning optical fiber story is another illustration of the value of research carried out by dedicated individuals with a clear vision of the future, when backed by management who shared that vision, and the resources both physical and financial that a large organization could provide.

LIGHT SOURCES AND DETECTORS
FOR OPTICAL FIBER SYSTEMS

An optical fiber communication system requires a light source that can be focused into a narrow beam for injection into the fiber and that is capable of high-bit-rate modulation, and a receiver, i.e., a photodetector, responsive to low-energy light. Ideally the light source is "coherent," i.e., it is confined to a single frequency and the phase is nonrandom; in practice there are many light sources with spectrum linewidths of much less than 1%. The coherence of the light emitted by the source is an important factor in concentrating the light into an intense narrow beam; it also facilitates efficient use of the spectrum, e.g., by high-bit-rate modulation and light carrier frequency multiplexing.

The invention of the laser (light amplification by stimulated emission of radiation) by A. L. Schawlow and C. H. Townes of Bell Laboratories in 1958, and the first realization of laser operation by T. H. Maiman of the Hughes Aircraft Company in 1960, provided an important first step.[4,5]

The basic operation of a laser consists of pumping atoms in a gas or solid into a stimulated state from which electrons can escape and fall to lower-energy ground states by giving up photons (light) of the appropriate energy. By providing such a device with a pair of parallel mirror surfaces several hundreds of odd-number half-wavelengths apart, continuous oscillations can be generated by an avalanche action. The pump energy may be from microwave, light, or other electromagnetic wave sources.

The helium-gas laser, operating at a wavelength of about 1 μm, was a useful near-coherent light source for laboratory experiments, but was hardly practicable for a communication system. The invention of the transistor stimulated the search for a semiconductor light source in Bell Laboratories and other laboratories during the 1960s. Early versions could only be operated at low temperatures and with intermittent operation. A breakthrough came in 1970 when Bell Labs first achieved continuous operation of a semiconductor laser at room temperature.[6] Although much development lay ahead, it was clear that a semiconductor device—of millimeter dimensions—could provide a near-coherent light source suitable for an optical fiber communication system.

Parallel research on light-emitting diodes (LEDs) produced a light source that, although not highly coherent (the linewidth at wavelengths of some 0.8 μm being typically about 1%), nevertheless could be formed into a sharply focused beam from a very small area. Moreover the LEDs could be readily modulated at bit rates of up to 10 Mbit/sec or more; they were thus well adapted to systems of modest capacity and length, operating with multimode fibers.[7]

One such device, with a single semiconductor junction, designed by C. A. Burrus of Bell Labs in 1971, emits light in a direction normal to the plane of the junction. Another "stripe edge" geometry results in

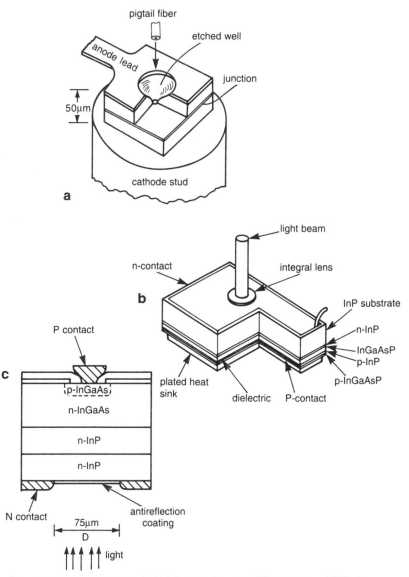

Figure 17.4. Light sources and photodetector for optical fiber systems. (a) The Burrus light-emitting diode (LED); (b) a 1.3-μm light wavelength LED; (c) a 1.3-μm light wavelength photodetector with transistor preamplifier.

light emission from the side of the junction. Yet another design—the "injection laser" diode—is similar to the Burrus diode but incorporates a gallium arsenide layer flanked by n- and p-type layers doped with indium phosphide. With this design it is possible to achieve an operating wavelength of about 1.3 μm which corresponds to a very-low-loss region of the light spectrum in optical fiber.[8a,9]

Development of light-sensitive photodiodes of the avalanche type has resulted in a modern Bell Labs design termed an "InGaAs PIN"; it incorporates a window with an antireflection layer, backed by an indium phosphide n-type photosensitive layer and an integral field-effect FET GaAs transistor preamplifier, designed for an operating wavelength of 1.3 μm.[8b]

The physics underlying the creation of the light sources and detectors that are as essential to optical communication systems as the fiber itself can be traced back to the atomic model delineated by J. J. Thomson and Ernest Rutherford, and extended by Niels Bohr (1885–1962). The further development of the quantum theory by Max Planck, Schröd-inger, Fermi, de Broglie, Niels Bohr, and others not only revealed much about the behavior of electrons, waves, and photons in the atomic world, it provided an indispensable basis for the design of optical devices (Chapter 2 and Ref. 10).

EARLY BELL SYSTEM OPTICAL FIBER SYSTEM FIELD TRIALS

During the 1970s the Bell Labs continued their development of long-life solid-state light sources and detectors, and optical fiber drawing techniques. In 1974 the latter underwent a significant advance with a chemical-vapor deposition technique, using a hollow tubular glass preform.[9] It will be recalled that Corning was also successfully using a vapor deposition technique inside a tubular glass structure.

A strong motivation for the first field trial of an optical system by Bell was to make better use of existing duct space to meet economically the ever-growing demand for interexchange telephone circuits in the local area. For such purposes the optical fiber system with its small size, low losses, and large circuit capacity offered an attractive solution.

The first field trial, carried out jointly by Western Electric and Bell Laboratories, was conducted at Atlanta, Georgia, in 1975.

The cable in this ambitious experiment carried 144 optical fibers with a loss of less than 6 dB/km, permitting a spacing of regenerative repeaters of at least 10 km. Each fiber operated at a bit rate of 45 Mbit/sec, using a GaAs laser light source of 0.82-μm wavelength. Thus, the complete cable had a potential telephone circuit capacity exceeding 100,000 telephone circuits.

During the period up to 1980, smaller cables with 24 fibers each operating at 45 Mbit/sec were developed and used operationally for telephone and video circuits. By February, 1983, an intercity trunk system operating at 90 Mbit/sec had been installed between Washington, D.C., and New York over a distance of 250 miles.

Thus, by the mid-1980s optical fiber systems had proved themselves both in the interexchange local area and for intercity trunks. The success of these inland systems paved the way, and gave confidence, for an attack on the next major objective—a transatlantic optical fiber cable system.

OPTICAL FIBER RESEARCH AND DEVELOPMENT IN THE UNITED KINGDOM

Early Work by the British Post Office

Interest in the possibilities of optical fibers for civil telecommunications, based mainly on theoretical studies, began in about 1965 at the Research Department of the British Post Office, Dollis Hill, NW London, and in the U.K. Government Signals Research and Development Establishment (SRDE) with a view to possible military applications. However, it was undoubtedly Charles Kao's work at STL and his classic paper of 1966,[1] outlined earlier in this chapter, that stimulated more detailed scientific studies and experimental work by the BPO at Dollis Hill. Thus began an investigation in depth of optical fiber systems that included the mechanisms of guided-light propagation in structured glass fibers and the losses therein, and the solid-state light sources and detectors needed for a communication system. BPO studies on optical fiber systems and devices up to the mid-1970s have been described in a series of papers in the *Post Office Electrical Engineers Journal*.[11-14]

The group at Dollis Hill initially responsible for this investigation was led by the scientists F. F. Roberts, J. I. Carasso, J. Midwinter, and

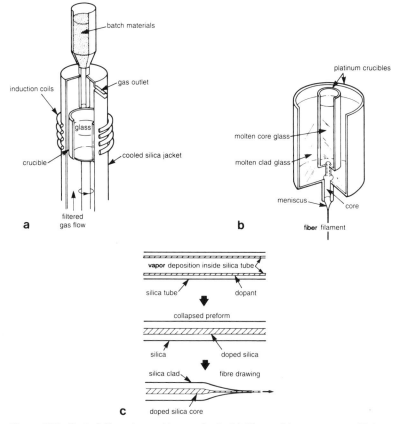

Figure 17.5. Optical fiber glassmaking methods. (a) Glassmaking apparatus; (b) double-crucible method; (c) vapor-deposition process for making silica-based optical fiber.

Dr. M. M. Faktor; their work was continued and expanded at the BPO Research Centre, Martlesham, Suffolk, opened in 1975 (now the British Telecom Research Laboratory). A member of this pioneering group, Dr. Marc Faktor (1930–1988), was awarded the BT Martlesham Medal in 1987 for outstanding scientific achievement and his pioneering work in optoelectronic techniques in the 1970s and 1980s. Much of his work was concerned with crystal growing from vapor, i.e., building up epitaxial layers of semiconductor material no more than a millionth of a centime-

Figure 17.6. Martlesham Medal winners for Achievement in Optical Fibre Transmission Systems. (Top) Dr. Marc Faktor, 1987 winner for pioneering work in optoelectronics; (bottom left) Dr. George Newns and (bottom right) Dr. Keith Beales, 1982 winners for innovation in optical fiber technology, including the double-crucible fiber manufacturing process.

ter thick—a process vital to the creation of light sources and detectors. In particular he pioneered a metal-organic vapor-phase epitaxy (MOVP) technique that was later exploited on a commercial scale jointly by BT and DuPont, in a worldwide market for optoelectronic components and devices valued at more than £4 billion. In the early 1970s he headed a team that used organic materials to build long-life semiconductor lasers that could be directly modulated. He also devised the electrochemical technique that is the heart of the BT "profile plotter" for displaying the electric-current carriers in semiconductor materials—a device now manufactured under license and sold worldwide.[15]

Marc Faktor was born in Poland in 1930 and came to the United Kingdom as a schoolboy in 1946, after grim wartime experiences. He began work in a London tutorial college and gained a first degree by evening class study at Birkbeck College, University of London, followed by a Ph.D. in inorganic chemistry in 1958, when he joined the BPO Research Department at Dollis Hill. He retired from BT in 1982, and became until his death a Visiting Professor at Queen Mary College, University of London.

He was a modest man of great personal charm who made, by his scientific skill and dedication, an enduring and major impact on the development of optical fiber communication systems.

RESEARCH, DEVELOPMENT, AND FIELD TRIALS OF OPTICAL FIBER SYSTEMS BY BRITISH TELECOM

The move of PO research from Dollis Hill to the new BT Research Centre at Martlesham, Suffolk, in 1975 with its improved facilities, and the challenges to a "privatized" BT from its competitors Mercury and Cable and Wireless, created both the capability and the incentive to pursue the development of optical fiber communication systems with determination and substantial resources. The continuing growth of telephone traffic and the introduction of new telecommunication services required increased capacity on junction and intercity trunk routes, and this had to be achieved at the lowest practicable cost. Similar requirements applied to the submarine cable routes between the United Kingdom and European countries, and across the Atlantic, to provide a balance with the growing capacity on satellite systems. And there were potentially important new applications for optical fiber systems in the local distribution network to customers' premises for cable television and enhanced telecommunication services.

There were several technological problems to be resolved, ranging from the need to ensure a large-scale and economic source of supply of the optimum design of stable, low-loss fiber cable, to the more practical problems of jointing cables in the field and inserting them in existing cable ducts.

Designs for vital components such as long-life light sources and optical detectors suitable for operation at high bit rates had to be established. And overall system engineering studies were needed to ensure the most economic design for each application. Economic and reliability considerations pointed firmly to the advantages of the lowest practicable fiber loss with the widest possible spacing of repeaters, and to monomode operation with its large communication capacity where this could be effectively utilized, e.g., on trunk routes. On the other hand, multimode fiber, with its larger core and lower manufacturing costs, might have advantages for smaller-capacity systems, e.g., in the local distribution network.

And the ultimate goal of a nonrepeatered transatlantic optical fiber submarine cable provided a challenging objective for all involved.

The following summary of BT research and development in optical fiber communication systems, part of which was carried out in collaboration with the U.K. telecommunications industry, outlines the more important technological milestones along the way and points to some of the pioneers whose achievements made these possible.

Low-Loss Optical Fiber Research

From earlier work in the United Kingdom, the United States, and elsewhere, two distinct optical fiber glass systems have emerged.

The one chosen for the U.K.-based effort, including the PO/BT laboratories, involved low-melting-temperature glasses such as the sodium calcium silicates (akin to window glass, but ultrapure) and the sodium borosilicates (of which Pyrex is one). These offer a wide range of optical, mechanical, and chemical properties; however, because they are melted in crucibles from powders, extreme care has to be taken to avoid impurities.

The other glass system, developed in the United States notably at Corning (see earlier in this chapter), uses pure fused silica (the glass form of quartz) and silica doped chemically to produce glass of a slightly different refractive index for the core material. The fused silica system has led to remarkably low losses—approaching 0.2 dB/km at 1.3-μm wavelength—which at the time of writing the low-melting-point glass method has yet to attain.

The pulling of molten glass into a fiber with the requisite structure and purity for low-loss light transmission, in unbroken lengths of 10 km or more, is a difficult and demanding task. One successful method uses the "double-crucible" process evolved and developed into a fiber manufacturing technique by two BT scientists, Dr. George Newns and Dr. Keith Beales. Their apparatus uses an inner platinum crucible to contain the molten core glass, and an outer crucible containing molten cladding glass with a lower refractive index. When electrically heated, molten glass begins to flow through concentric nozzles and can be pulled into a filament wound onto a drum as it cools and solidifies. By careful control of temperature distribution and winding speed, the correct dimensions of core and cladding can be achieved. By choice of nozzle dimensions, either mono- or multimode fiber can be made; the process is particularly well suited to the low-cost production of multimode fiber with its relatively large inner core. It also enables graded-index fibers to be produced, the latter being achieved by a diffusion process between core and cladding before solidification occurs. The Japanese "Selfoc" fiber is made in this manner. The two-crucible fiber production process originated by George Newns and Keith Beales has been licensed to British, American, and European companies. Newns and Beales's later work has included further development of the vapor-deposition process using silica glass preforms, similar to the Corning process described earlier in this chapter.

For their outstanding contributions to glass fiber technology, Dr. George Newns and Dr. Keith Beales were awarded the Martlesham Medal in July, 1982.[16]

George Newns, who was born in 1936, gained a first-class honors B.Sc. degree at Liverpool University, followed by a Ph.D. He joined the BPO in 1962, and after working on semiconductors he was appointed in 1968 to lead a research team investigating optical materials, and later fiber fabrication and evaluation for optical communications. He has published over 50 scientific papers and holds a number of patents. With

Dr. Beales and other colleagues he was awarded the Institution of Electrical Engineers (London) Scientific Premium for a paper on the double-crucible fiber process, and in 1980 the Potts Medal by the Chemical Society of Liverpool University.

Keith Beales, who was born in 1940, won an Open Scholarship to Nottingham University where he gained an honors B.Sc. degree in chemistry and a Ph.D. for a thesis on organic semiconductors. He joined the Research Department of the BPO in 1966 where his work involved the materials aspects of semiconductor devices and optical fibers.

Since 1979 he led the research and development on optical fibers, including the double-crucible process, monomode fiber development, and fiber strength and reliability studies. He has published over 50 scientific and conference papers and filed 15 patents.

Mention should also be made of the valuable role played by the Optoelectronics Research Centre at the University of Southampton under Professor W. A. Gambling. The Southampton team developed in the 1970s a process for making low-loss fiber based on phosphorus-doped silica, using technology similar to that invented at AT&T/Bell by J. MacChesney (modified chemical-vapor deposition). Much good work on optical fiber assessment and propagation studies has since been carried out over the years at Southampton University, including the development of special fibers for sensors.

To British Telecom Research at Martlesham should go the credit for their early recognition of the longer-term advantages of single-mode, as opposed to multimode, fiber; their work on the development of single-mode fiber was acknowledged by the Queen's Award to the fiber section of BT Research in 1985.

British Telecom Optical System Field Trials

By the early 1970s the development of optical fibers and devices needed for a communication system had progressed to a point where field trials could be envisaged. The emphasis was initially on the use of multimode fiber operating at a wavelength of about 0.8–0.9 μm and which, by virtue of its relatively large core diameter, could be more efficiently coupled to LED light sources and to detectors. It was also more readily spliced (joined) under field conditions than monomode fibers.

At the opening of the British Telecom Research Centre at Martlesham in 1975, an 8-Mbit/sec 6-km laboratory-based optical fiber link was successfully demonstrated using multimode fiber at a wavelength of about 0.8 μm.

In 1977 field trials using graded-index multimode fiber operating at 140 Mbit/sec and installed in ducts in the Martlesham area gave the first indications that fiber cables could be installed in ducts, spliced where required, and operated over repeater sections longer than on conventional copper cables. The 140-Mbit/sec rate meant that a single fiber could provide some 2000 telephone channels or a high-quality television channel.

By the late 1970s the scene was changing dramatically. Telecommunication research laboratories worldwide were drawing attention to the lower fiber losses to be achieved by operating at wavelengths of about 1.3 or 1.5 μm, and the superior communication capacity offered by monomode fibers. Optical device and component research and development began to be diverted to this new region of the light spectrum.

In 1980 BT Research Laboratories demonstrated a 140-Mbit/sec 1.3-μm-wavelength optical fiber link using components developed at BTRL and operating over an unrepeatered 37-km length of monomode silica cable. This was a crucial demonstration since it meant that the power-feeding sections in the existing BT trunk network, spaced up to 30 km, could now be bridged by an unrepeatered section of optical fiber cable—a result that gave a substantial economic advantage to the optical fiber system compared with conventional coaxial cables.[17]

Also in 1980 a 10-km submarine cable link, using components identical to the land system referred to earlier, was demonstrated at Loch Fyne, Scotland, in collaboration with Standard Telecommunication Laboratories.

During 1982 two significant field trials of duct-installed single-mode optical fiber cable operating at a wavelength of 1.3 μm were carried out by BT in Suffolk, partly with a view to examining the effect of cable tolerances on splicing losses. The first, carried out in cooperation with Telephone Cables Ltd., involved the installation of a 7.5-km route near Woodbridge, Suffolk, using experimental fiber made by BTRL and GEC Optical Fibres Ltd. The second, made in association with STL, was over a 15-km route between Martlesham and Ipswich using cable made by STL/STC in a pilot manufacturing plant. The splicing loss tests revealed clearly the benefits of the closer dimensional tolerances of the second cable. The Woodbridge cable was operated initially at 140 Mbit/sec and later at 650 Mbit/sec; the Martlesham–Ipswich cable was operated at 565 Mbit/sec over a looped distance of 61.3 km, an achievement believed to be the first of its kind.

Additional tests at a wavelength of 1.55 μm achieved a bit rate of 140 Mbit/sec over an unrepeatered looped distance of 90.5 km on installed Martlesham–Ipswich cable.[17]

These field trials created a sound basis for proceeding in the United Kingdom with a nationwide junction and trunk network based on monoptical fiber systems; they also created confidence that there would portant role for optical fiber submarine cables, even spanning the Atlantic Ocean.

By 1982 British Telecom had several thousand kilometers of first-generation 0.8-μm-wavelength optical fiber cable systems in service. In July, 1982, a second-generation 1.3-μm-wavelength 140-Mbit/sec, repeatered optical fiber link 204 km long—then the world's longest optical link—went into service between London and Birmingham. This link used cable made by BICC Telecommunication Cables Ltd. and equipment made by Plessey Telecommunications Ltd. to BT specifications.[18]

And a new name was coined by British Telecom for its optical fiber cable systems—"Lightlines."

By 1986 there were more than 65,000 km of second-generation 140-Mbit/sec optical fiber cable systems in use in the BT network. In that year two 565-Mbit/sec optical fiber systems were also put into service—one of which used ultrareliable optoelectronic and microchip components designed and made by BTRL as a test-bed for component designs to be used in a transatlantic optical fiber cable (TAT 8).[18]

Optical Fiber Cable Systems Go under the Ocean

Following the successful optical fiber submarine cable experiment in Loch Fyne in 1982, BT initiated plans for operational links across the North Sea and to Ireland.

The world's first international optical fiber submarine cable link, between the United Kingdom and Belgium, was opened in October, 1986, with a two-way London–Ostend video conference. The cable system was supplied by STC Submarine Cables Ltd. to BT order and specification, the main 80-km deep-water section being laid by the British Telecom International cable ship *Alert*. The optical fiber system provides more than 11,500 additional telephone circuits across the North Sea.

In 1988 optical fiber cables were laid across the Irish Sea, one 90 km long between Anglesey and the Isle of Man, and another 126 km long between Anglesey and Portmarnock in the Irish Republic. Neither of these cables requires intermediate repeaters on the seabed—a clear illustration of the remarkable success achieved by the long scientific and engineering battle to reduce losses in optical fibers. They paved the way for an even more dramatic success in the optical fiber story—the spanning of the Atlantic Ocean by an optical fiber cable system.

A comprehensive collection of papers on optical fiber submarine cable system development has appeared in the *Journal of the Institution of British Telecommunication Engineers.*[19]

THE FIRST TRANSATLANTIC OPTICAL FIBER SUBMARINE CABLE SYSTEM (TAT 8)

The world's first transatlantic optical fiber cable system, TAT 8, capable of carrying 40,000 simultaneous telephone conversations, was opened for service between Europe and North America on December 14, 1988, by AT&T, British Telecom, and France Telecom. TAT 8 effectively doubled the existing cable capacity across the North Atlantic and, in addition to large-scale telephone service, it transmits data and video signals.[20]

TAT 8 was a major telecommunication project remarkable for the degree of international cooperation in the design, management, financing, and operation of the enterprise. Ownership is vested in AT&T 34%, BT 15.5%, and FT 10%, the remainder being taken up by other telecommunication administrations. British Telecom and its industrial partners, notably STC, played a very important role in planning the new system through its earlier experience in submarine cable systems and the technological support available from its Research Laboratories.

From the landing point at Tuckerton, New Jersey, the cable extends 5800 km across the Atlantic to a branching point off the coasts of the United Kingdom and France on the European Continental Shelf; from

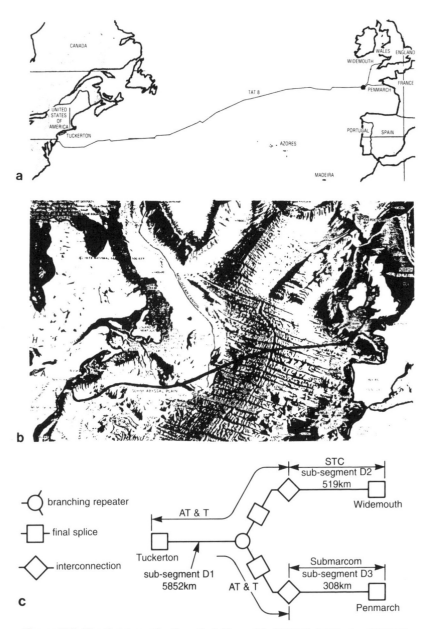

Figure 17.7. The first transatlantic optical fiber cable (TAT 8). (a) Route of TAT 8 cable; (b) bed of Atlantic Ocean; (c) participants and cable provision plan.

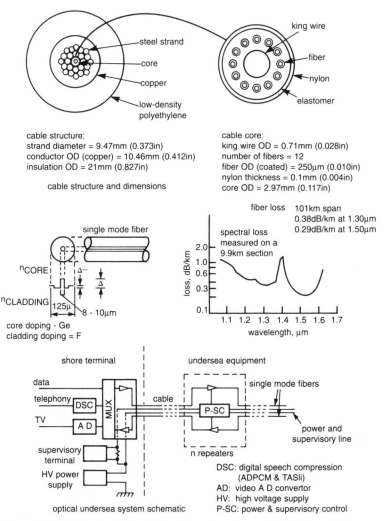

cable structure:
strand diameter = 9.47mm (0.373in)
conductor OD (copper) = 10.46mm (0.412in)
insulation OD = 21mm (0.827in)

cable structure and dimensions

cable core:
king wire OD = 0.71mm (0.028in)
number of fibers = 12
fiber OD (coated) = 250µm (0.010in)
nylon thickness = 0.1mm (0.004in)
core OD = 2.97mm (0.117in)

fiber loss   101km span
0.38dB/km at 1.30µm
0.29dB/km at 1.50µm

spectral loss
measured on a
9.9km section

core doping - Ge
cladding doping = F

DSC: digital speech compression
(ADPCM & TASli)
AD: video A D convertor
HV: high voltage supply
P-SC: power & supervisory control

optical undersea system schematic

Figure 17.8. TAT 8 optical fiber cable: cable and equipment details. (From *Fiberoptics*, TAB Books.)

this point the U.K. branch continues 540 km to Widemouth, N. Devon, and the French branch 330 km to Penmarch, Brittany.

The TAT 8 system provides two pairs of fibers, one pair routed from Tuckerton to Widemouth, United Kingdom, and the second pair from Tuckerton to Penmarch, France. The two pairs separate at a branching repeater on the European Continental Shelf and, to keep symmetry within the cable and facilitate restoration, a pair is provided between the United Kingdom and France. The branching repeater provides for switching the fibers between the different locations for operational security reasons. Each fiber pair carries two both-way 140-Mbit/sec signals.

The main transatlantic section of the cable system from the United States up to and including the branching repeater has been provided by AT&T, the section to the United Kingdom by STC Submarine Cables Ltd., and the section to France by Submarcom. The sections provided by the various manufacturers differ in design to some degree, e.g., in the standby arrangements to secure high reliability. However, all are based on the use of monomode fiber operating at a wavelength of about 1.3 µm.

The mechanical aspects of cable design differ between the three suppliers; each provides a home for the optical fibers along the neutral axis of the cable but uses different techniques to ensure mechanical strength and protection against water and hydrogen access under the extreme pressures encountered in the depths of the ocean. (Glass fibers are susceptible to deterioration in the presence of heavy concentrations of free hydrogen.) Heavy armoring is provided to guard against damage, e.g., by trawlers, in the shallow-water sections. A plough, designed by the BT Research Laboratories, has been used to plow in cable in shallow waters to provide additional protection against such damage. The cables were laid by the cable ships of the cooperating administrations, the U.K. section being laid by the BT cable ship *Alert*. The comprehensive telemetry facilities used to locate and correct faults and the arrangements for maintenance of the cable system are described in Ref. 19.

The success of the pioneering TAT 8 cable, the massive communication capacity it provides, and the ability of digital optical fiber systems with branching repeaters to serve multiple destinations, together with the operational flexibility and economy this confers, have greatly broadened the scope for such systems.

Satellites now have a very worthy competitor for providing worldwide telecommunication services!

Light Can Now Be Amplified, Multiplexed, and Switched in Optical Fibers

By the late 1980s some important new developments were beginning to emerge from British Telecom, AT&T/Bell, and other research laboratories investigating optoelectronic techniques; these included:

- The direct amplification of light in a fiber
- Multiplexing light of different wavelengths in a single fiber
- Switching light from one fiber to another
- Generating light with a high degree of coherence

Although the amplification of light in a solid-state laser is possible, the most promising light amplifier to date is one using stimulated emission in, or with, a conventional silica-based monomode fiber, e.g., operating at 1.55 µm, doped with low concentrations of the rare-earth ion erbium, and pumped by light of a different wavelength, e.g., 0.98 µm. Such light amplifiers have very low inherent noise and can offer gains of 10 dB or more with power outputs of up to about 1 watt. Moreover, a relatively wide bandwidth is possible, permitting operation up to at least 2.5 Gbit/sec (2500 Mbit/sec) and opening the door to multiplexing light of different wavelengths on a single fiber with a common amplifier.[21]

Early pioneering work on light amplification in neodymium-doped glass fiber lasers operating at 1 µm was carried out by E. Snitzer and C. J. Koestler of the American Optical Company, Massachusetts, in the 1960s.[22] A very important step forward was carried out by D. N. Payne and his colleagues at Southampton University in the mid-1980s; they demonstrated some 26-dB gain in erbium-doped single-mode glass fibers operating at 1.5 µm, pumped at 0.65 µm.[23]

The practical implications of this advance are substantial: existing monomode fiber cables in the inland junction and trunk networks operating at 140 Mbit/sec can be upgraded in capacity by factors of ten or more by replacement of repeaters; repeater spacings can be greatly increased.

A potential application also exists in future local distribution optical fiber networks to customers' premises, as a wideband power distribution amplifier handling several multiplexed signals on different light wavelengths.

The low-noise property has already found an important application in submarine optical fiber systems as a preamplifier to a conventional

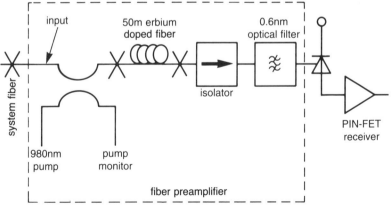

system fiber

input

50m erbium
doped fiber

0.6nm
optical filter

isolator

980nm
pump

pump
monitor

fiber preamplifier

PIN-FET
receiver

light detector, providing a valuable improvement in sensitivity and enabling a wider spacing of repeaters to be achieved.

A team at the British Telecom Research Labs, led by Dr. Derek King, Alec Parker, and Richard Calton, was responsible for the design and manufacture in quantity of an optical receiver for submarine systems that included a light preamplifier and achieved the highest sensitivity then available anywhere in the world.

The high-sensitivity optical receiver has been used in TAT 8 and numerous submarine systems in the North Sea, Mediterranean, and Pacific Ocean. In recognition of this outstanding work, BTRL was awarded the prestigious Queen's Award for Technological Achievement in 1990.[24]

"Blown Fiber"

Not all of the developments in the optical fiber world have been on the plane of high technology: an innovation of great practical and economic importance has been the development of the "blown fiber" technique for installing optical fiber cables in customers' premises and under pavements in the local network. The speed and low cost of fiber installation by this method have made an outstanding contribution to the spread of optical networks.

In this technique bundles of three or four optical fibers, held together in a polyethylene sleeve, are blown down a small tube or "microduct." The microduct is a tough and flexible tube with simple "push-fit" connector joints; using compressed air the bundle of optical fibers can be blown distances of 2 km or more through the microduct.

The development was carried out at BTRL by Mick Reeves and his colleague Steven Cassidy and led to field trials in 1984. In 1989 the Martlesham Medal was awarded to Mick Reeves. The citation for the award noted that "Blown Fibre is a fine example of applied research of the very best sort. Thanks to Mick's dedication and imagination it has been refined from his original idea into a practicable, marketable product."

Figure 17.9. British Telecom Research Laboratories at Martlesham were awarded the Queen's Award for Technological Achievement (optoelectronic detectors) in 1990. (Top) British Telecom team responsible for the design and production of photodetectors for submarine cable systems (left to right: Dr. Derek King, Alec Parker, and Richard Calton). (Bottom) Low-noise light amplifier using erbium-doped fiber.

REFERENCES

1. K. C. Kao and G. A. Hockham, "Dielectric Fibre Surface Waveguides for Optical Frequencies," *Proc. IEE, 113*, No. 7, July 1966.
2. I. C. Magaziner and M. Patinkin, *The Silent War* (Random House, New York, 1989), Chapter 9.
3. R. D. Maurer, "Glass Fibers for Optical Communications," *Proc. IEEE, 61*, April 1973.
4. A. L. Schawlow and C. H. Townes, "Infrared and Optical Measures," *Phys. Rev., 112*, December 1958.
5. T. H. Maiman, "Stimulated Optical Radiation in Ruby," *Nature, 187*, August 1960.
6. M. B. Panish, "Heterostructure Injection Lasers," *Bell Lab. Rec., 49*, November 1971.
7. C. A. Burrus and R. W. Dawson, "Small-Area High-Current-Density GaAs Electroluminescent Diodes and a Method of Operation for Improved Degradation Characteristics," *Appl. Phys. Lett., 17*, August 1970.
8. J. A. Kuecken, *Fiberoptics: A Revolution in Communications* (TAB Books, Blue Ridge Summit, PA, 1987), (a) Chapter 15, (b) Chapter 14.
9. *A History of Engineering and Science in the Bell System: Transmission Technology (1925–1975)*, edited by E. F. O'Neill (Bell Laboratories, 1985), Chapter 20.
10. T. Hey and P. Walters, *The Quantum Universe* (Cambridge University Press, London, 1987).
11. K. J. Beales, J. E. Midwinter, G. R. Newns, and C. R. Day, "Materials and Fibres for Optical Transmission Systems," *Post Off. Electr. Eng. J., 67*, July 1974.
12. R. B. Dyott, "The Optical Fibre as a Transmission Line," *Post Off. Electr. Eng. J., 67*, October 1974.
13. D. H. Newman, "Sources for Optical Fibre Transmission Systems," *Post Off. Electr. Eng. J., 67*, January 1975.
14. M. R. Mathew and D. R. Smith, "Photo-Diodes for Optical Fibre Transmission Systems," *Post Off. Electr. Eng. J., 67*, January 1974.
15. "Martlesham Medal for Pioneer (Dr. M. M. Faktor) in Opto-Electronic Technology," *Br. Telecom. Eng., 6*, April 1987.
16. "Martlesham Medal for Optical-Fibre Pioneers" (Dr. G. Newns and Dr. K. J. Beales), *Br. Telecom Eng., 1*, October 1982.
17. R. C. Hooper, "The Development of Single-Mode Optical Fibre Transmission Systems at BTRL," *Br. Telecom Eng., 4*, July 1985.
18. "565 Mbit/sec Optical Fibre Link," *Br. Telecom Eng., 5*, April 1986.
19. "Optical-Fibre Submarine Cable Systems," *J. Inst. Br. Telecom Eng., 5*, Pt. 2, July 1986.
20. "TAT 8 Goes Live," *Br. Telecom Eng., 7*, January 1989.

21. *British Telecom Technical Review*, Autumn 1989 and Spring, Summer 1990.
22. C. J. Koestler and E. Snitzer, "Amplification in a Fiber Laser," *J. Appl. Opt.*, 3, No. 10, October, 1964.
23. R. J. Mears, L. Reekie, I. M. Jauncy, and D. N. Payne, "High-Gain Rare-Earth-Doped Fibre Amplifier at 1.54 Micron," Department of Electronics and Information Engineering, Southampton University, 1987.
24. "Labs Win Fourth Queen's Award," *British Telecom Technical Review*, Summer 1990.

18

Inventors of the Visual Telecommunication Systems

The invention and development of the telephone enabled the simplest and most direct form of human communication to be achieved, that is, by voice and ear, but left unused another human sense organ, namely the eye. There was clearly scope for a variety of modes of visual communication, ranging from the transmission of hand- or typewritten documents, sketches, and photographs (facsimile), to live video conferencing between pairs or groups of people (teleconferencing), and teletext (the presentation of alphanumeric and other information on television receivers or video display units).

FACSIMILE TRANSMISSION*

The early history of facsimile transmission goes back to 1842 when the Scottish inventor Alexander Bain developed a pendulum system

---

*This account is based largely on Daniel Costigan's book about facsimile.[1]

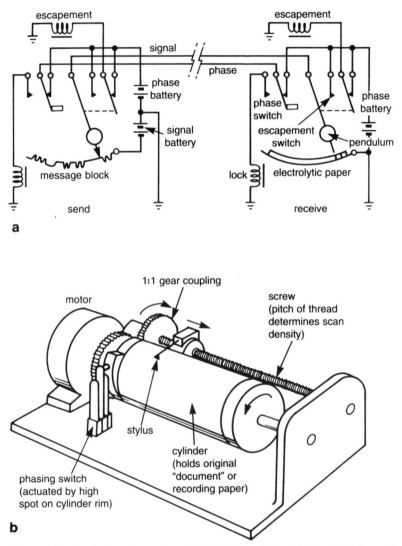

a

b

Figure 18.1. Facsimile systems (from D. M. Costigan's book *Electronic Delivery of Documents and Graphics*[1]). (a) Alexander Bain's "electro-chemical recording telegraph" (1843); (b) Frederick Bakewell's "cylinder-and-screw scanning system" (c. 1850); (c) modern facsimile machine for whole newspaper pages (Muirhead Systems Ltd.).

Figure 18.1. (Continued)

for synchronizing a master–slave system of clocks. He realized that if the master pendulum traced a series of electric contacts, e.g., corresponding to letters of the alphabet, a slave pendulum with a stylus carrying an electric current could reproduce these signals on chemically impregnated paper—this he called "an electro-chemical recording telegraph," for which he was granted British Patent No. 9745 in 1843. The first commercial system was established in France in 1865, linking Paris with several cities and using a modification of Bain's device patented by Giovanni Caselli.

Bain's system was later superseded by electromechanical printing telegraph systems such as that invented by David Hughes in 1854 (see Chapter 3). Nevertheless, by introducing the concepts of synchronization and the creation of an image from a varying electric current, Bain had laid the basic foundations of modern facsimile systems.

A major step forward in the evolution of facsimile systems was made in 1850 by the Englishman Frederick Bakewell, who demonstrated the "cylinder-and-screw" arrangement that replaced Bain's pendulum and is still used in some facsimile systems today.

The next important step was the use of the light-sensitive element selenium, first suggested by Shelford Bidwell in England, as a means for converting a scanned image into a varying electric current—a technique that opened the door to "halftone" transmission, i.e., of intermediate tones between white and black. The first practical photoelectric facsimile system was demonstrated in 1902 by the German inventor Dr. Arthur Korn.

Commercial use of this system began in Germany in 1907, and by 1910 Berlin, Paris, and London were linked by facsimile transmission over the telephone network. By 1922 Korn's system had been used to transmit by radio a photograph of Pope Pius XI from Rome to Maine, USA, the received picture being published the same day in the *New York World* newspaper. Thus began what was to become one of the most valuable uses of facsimile transmission—by the press, initially for individual photographs of up-to-the-minute news, but eventually for whole pages of newspapers from news-gathering offices in cities to remote presses.

Following Korn's successful demonstrations, facsimile transmission systems—the term is sometimes shortened to *fax*—began to be developed commercially by the Marconi Co. and by Muirhead in England, Telefunken in Germany, and Belin in France. In the United States AT&T supported by pioneering research in the Bell Labs,[2] the Radio Corporation of America, and the Western Union Telegraph Co. set about developing picture transmission systems, initially for press use.

One of the technically most interesting facsimile systems, developed in England in 1920, two years before Korn's transatlantic radio triumph, by H. G. Bartholomew and M. L. MacFarlane and called the "Bartlane" system, enabled picture transmissions to be made over the transatlantic telegraph cable (see Chapter 3). It converted picture elements to groups of perforations in paper tape in a manner anticipating present-day pulse-code modulation (PCM) digital systems. The perforated paper tape was used to transmit on–off signals over the cable to the receiving end where each group of coded signals re-created a picture element. Clearly, because of the restricted bandwidth available on the cable, this was a very slow process.

In addition to the press usage of facsimile, RCA pioneered its use for the transmission by radio of weather maps to ships at sea, from which developed a major usage in weather forecasting. Some of the early facsimile workers were seized with the idea of using radio broadcasting transmitters to transmit newspapers, either during the night hours when

the transmitters were not in use for sound broadcasting, or on an ampli-
tude- or frequency-modulated tone higher in frequency than the sound
channel. One fax system called the "Fultograph" after its Austrian
inventor Captain Otho Fulton was operated experimentally for a few
years beginning in 1928, in at least four European cities, including Berlin.
Several fax newspaper systems were tried out on a semicommercial
basis in the United States during the 1930s and 1940s, and by 1948 the
U.S. Federal Communications Commission had authorized commercial
fax broadcasting. But by then television broadcasting had made its debut
and was commanding public attention; fax broadcasting virtually ceased
but fax continued to develop in its more utilitarian roles for press,
weather map distribution, and business users.

In the early 1930s a number of developments took place that later
significantly improved the quality, speed, and convenience of facsimile
transmission. One was the all-electronic flying-spot scanner invented
by Manfred Von Ardenne in Germany , and the other a modulated ink-
jet recording method developed by C. W. Hansell of RCA. John Logie
Baird in England (see Chapter 9) was investigating in 1944 the possibility
of using his television scanning techniques for high-speed facsimile
transmission in wide band radio channels. Finch in the United States
demonstrated a system of color fax, a high-speed cathode-ray tube
scanner, and, in 1938, the simultaneous transmission of fax and high-
quality sound on FM VHF broadcasting transmitters.

The development of television spurred on other electronic device
developments that found application in facsimile systems, including
the photomultipliers pioneered by Zworykin and Farnsworth. The facsimile art itself continued to develop from the 1950s onwards, notable
advances being electrostatic recording (xerography, RCA's "Electro-
fax"), improved scanning by lasers, and Bell Labs' charge-coupled devices (CCDs).

Transmission was speeded by techniques that reduced redundancy,
e.g., by fast scanning over white spaces between letters and words,
evolved by Kretzmer, Hochman, and others. By the late 1970s the microchip and the microprocessor had become available (see Chapter 12);
this simplified and improved the operation of fax equipment, made it
more flexible, and at the same time contributed substantially to reliability and economy.

These developments, which enabled the large-scale manufacture
of compact, low-cost fax equipment, opened the door to a vast new
market for business and personal use. By the 1980s facsimile was an

accepted part of the armory of information technology that is transforming the business world.

Vitally important for the growth of the facsimile transmission market was international standardization of the technical characteristics and operating procedures through the International Telephone and Telegraph Consultative Committee (CCITT), initially directed to the use of 3-kHz-bandwidth telephone channels. At first, in the 1960s, it took six minutes to transmit an A4 page (Group 1 standard). By 1976 the Group 2 standard had reduced this to three minutes and improved the quality with a scanning density of 100 lines per inch. Further developments followed with Group 3 machines, which use digital encoding and transmit an A4 page in less than one minute over a telephone channel, with high-quality reproduction of text and diagrams.

A new generation of facsimile machines—Group 4 standard—is set to surpass the Group 3 machines as the Integrated Services Digital Network (ISDN) becomes established on a global basis in the 1990s. By exploiting the 64-kbit/sec capability of the ISDN the transmission time of an A4 page is reduced to less than 30 seconds, with faster connection setup time between sender and receiver.

The provision of hard copy of documents by facsimile techniques further enhances the visual communication capabilities of the ISDN which include person-to-person teleconferencing and VDU document display (Chapter 19).

TELEVISION CONFERENCING

From the 1950s onwards an increasing effort was being put into research aimed at providing a live visual element to person-to-person and group-to-group communication at a distance, with the objective of making such communication as close to direct, in-the-same-room contact as was practically and economically possible.

The Analogue Videophone

The first efforts toward this objective fell far short of the ideal—they took the form of the "Picturephone" developed by Bell Laboratories in the United States[3] and a similar development in the United Kingdom,

Figure 18.2. British Telecom "Video-phone" (1994).

the British Post Office "Viewphone." Both offered the telephone user a head and shoulders picture of the other party and limited capability for displaying alphanumeric data. The Picturephone used a 251-line, 30-frames/sec picture requiring a video bandwidth of 1 MHz; the Viewphone offered a 319-line, 25-frames/sec picture in a video bandwidth of 1 MHz. Even with this restricted bandwidth the links to the customers' premises and for intercity transmission were complex and costly: the Picturephone required three cable pairs to each customer's premises, one pair for voice and two pairs for the go and return video signals.

Even when the video signal was encoded to the lowest practicable rate of 6.3 Mbit/sec for intercity transmission, only three long-distance circuits could be provided in the 20-MHz baseband of microwave radio-relay systems. Extensive AT&T/Bell System field trials of Picturephone service were carried out in 1965–1966 between New York and Chicago but failed to attract customers on a sufficient scale to justify further development of the service.

But by the early 1990s prospects for the videophone had improved dramatically. The ongoing development of digital and coding methods had opened the door to new techniques for drastically reducing the bit rate by discrete cosine transform (DCT). This made it possible to transmit acceptable-quality color video signals, albeit at the relatively low picture rate of 8 per second, over customers' conventional analogue telephone lines. By 1994 both AT&T and British Telecom were offering videophone service based on the new digital technology, although at different bit rates (AT&T 19.2 kbit/sec, BT 14.4 kbit/sec).

The Higher-Definition Digital Videophone

With the advent of the ISDN a higher-definition videophone becomes possible. In a British Telecom development, a 64-kbit/sec channel in the ISDN is used, with advanced coding of the video signal and a picture repetition rate of 15 pictures/sec, to provide picture quality approaching that of normal broadcast television. The picture can be displayed on a conventional television monitor or as a "window" on a computer VDU; in the latter case it has value for sales and shopping services as well as providing person-to-person visual contact.

Group-to-Group Teleconferencing

Television conferencing or "teleconferencing" can provide a service of great value for business users and large organizations with dispersed office/factory locations. The ultimate objective is to achieve electronically the same ease of communication between groups of people separated by distance as if they were in the same room. The value of teleconferencing to the users increases with distance because the costs of travel and staff time are themselves distance related. It is important therefore to minimize the costs of long-distance transmission, as well as the link to the users' premises, if a teleconferencing service is to be economically attractive.

For maximum effectiveness of communication a group-to-group television conference facility requires a higher-definition and larger picture than is available in the videophone services; a natural first choice was the television broadcasting standards of 525 lines in the United States and 625 lines in the United Kingdom and Europe for which

PUNCH'S ALMANACK FOR 1879.   [December 9, 1878]

EDISON'S TELEPHONOSCOPE (TRANSMITS LIGHT AS WELL AS SOUND).

(*Every evening, before going to bed, Pater- and Materfamilias set up an electric camera-obscura over their bedroom mantel-piece, and gladden their eyes with the sight of their Children at the Antipodes, and converse gaily with them through the wire.*)

Paterfamilias (*in Wilton Place*). " BEATRICE, COME CLOSER, I WANT TO WHISPER."   *Beatrice* (*from Ceylon*). " YES, PAPA DEAR."
Paterfamilias. " WHO IS THAT CHARMING YOUNG LADY PLAYING ON CHARLIE'S SIDE?"
Beatrice. " SHE'S JUST COME OVER FROM ENGLAND, PAPA.   I'LL INTRODUCE YOU TO HER AS SOON AS THE GAME'S OVER!"

Figure 18.3. Mr. Punch foresaw conference television 100 years ago. It is remarkable that this cartoon in *Punch* magazine for December 1879 foresaw not only U.K.–Australia conference television but also large-screen projection in a format not unlike today's "high-definition" TV!

cameras, television receivers, and intercity transmission facilities are readily available.

In the early 1970s the BPO Research Department carried out experimental field trials of a television conference service using modified broadcast standard television equipment, enabling groups of up to about ten people to communicate visually as well as aurally. The research studies included means for video bandwidth reduction by optimizing picture format for the conference service and reducing redundancy in the picture signal; two conference video channels could thus be provided in a broadcast standard intercity video link.

In 1972 the BPO opened its "Confravision" service, with studios in London and four other U.K. cities for hire by business and other users, claimed as "a world-first purpose-built system of conferences by television."[4] The studios were also equipped with facilities for dis-

Figure 18.4. BT "split-screen" Confravision studio.

playing documents on a television screen and high-speed facsimile transmission, virtually enabling copies of documents to be "handed across the table" as in a direct conference situation.

The long-distance capabilities of Confravision were further demonstrated in 1979 with the setting up of Europe's first conference by satellite between London and the Telecom '79 International Conference in Geneva.[5] And in 1984 the Ford Motor Company and British Telecom International set up a Confravision link between studios in its Essex plant and another in Cologne, West Germany, via the European Communication Satellite ECS1, for daily consultations between its staff at the two centers.[6]

In the 1980s, television broadcasters began to use television conferencing as a ready and effective means for conducting interviews, generally on a "one-to-one" basis, and often on a worldwide scale via communication satellites. There was also the occasional use of multiple-screen,

or split-screen, techniques for conducting discussions involving participants in three or more distant locations simultaneously.

It has to be admitted that up to the present time (early 1990s) the
use of teleconferencing by business and other organizations has been
on a limited scale, in spite of the manifest advantages of immediacy,
convenience, and time saving it can offer.

Developments that are opening the door to larger-scale use of
teleconferencing in the future are:

- Lower costs for local area, intercity, and international transmission
- Ability to "dial up" a video link as readily as a telephone circuit

Teleconferencing and the Integrated Services Digital Network

The development of an integrated services digital network (Chapter
19) will enable video conferencing to be achieved on a worldwide scale
at viable cost using transmission and switching facilities common to
telephone, facsimile, and other services. The European collaborative
project MIAS (Multipoint Multimedia Conference System) enables groups
of users at different locations to see and speak to one another, send
facsimile documents, exchange data files, and examine still-image presentations.[9]

Two teleconferencing modes using the ISDN are offered:

- The "videophone," referred to above, using a 64-kbit/sec digital
  channel, providing picture quality adequate for person-to-person
  communication and telephone speech
- A "videoconferencing" mode at 2 Mbit/sec with higher-quality
  pictures suitable for group-to-group communication

The success of this development depends heavily on advanced designs
of "codecs"—a codec (signal coder/decoder) converts analogue signals
for speech, facsimile, or pictures into digital form for transmission over
a digital path and reconverts them at the receiving end. By using sophisticated coding algorithms the bit rate needed for desired speech or
picture quality at the receiving end can be substantially reduced. It is the
development of this technique, in which members of the BT Martlesham
Laboratories played a prominent part, that enables the videophone to
overcome the practical and economic limitations of the Bell Picturephone

and Post Office Viewphone referred to earlier in this chapter. Furthermore, the ISDN also provides the key to readily dialing up videophone and teleconferencing connections.[9]

The Future of Teleconferencing

Optical fiber systems, with their wide available bandwidth and low losses (Chapter 17), leading to lower costs per megabit per second per kilometer, have very important roles to play in both long-distance transmission and the local network. There are roles too for millimetric radio systems, both in the local network and for providing additional capacity in communication satellites.

There is scope for improving the communication effectiveness of group-to-group television conferencing from the user viewpoint by larger screens with higher definition—to which current developments in 1250-line television broadcast systems using more sophisticated coding may point the way.

There is also the as yet inadequately explored field of three-dimensional presentation, with stereophonic sound. The single camera and microphone can only represent the eyes and ears of one of a dispersed group of people; realism requires that each participant sees and hears the distant group from his or her distinctive viewpoint.

Bearing in mind the very substantial benefits that effective and economic teleconferencing could bring to the business world, educational, political, social, and other organizations, and the environmental advantages of reducing travel, it is clear that there is ample scope in this field for future innovators and invention.

VIEWDATA (PRESTEL) AND TELETEXT (CEEFAX AND ORACLE)

While the aural presentation of recorded information, e.g., in the telephone system for instructional purposes, has a certain personal feel, visual presentation has significant advantages:

- Information can be presented at a rate closely related to the recipient's ability to absorb it since it can be held on the screen for as long as is needed.

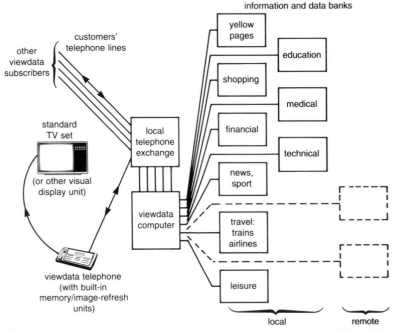

Figure 18.5. Principles of PO "Viewdata" Visual Information Service (1974). This later became British Telecom "Prestel" service.

- It can include graphic information.
- It can be more readily made "interactive," i.e., responsive to user requirements.
- It lends itself to permanent record form, e.g., on the printed page.

These considerations led in the 1970s to an active, but initially independent, investigation of the possibilities of including a visual information service in the BPO telecommunication system, and by the British Broadcasting Corporation and the Independent Broadcasting Authority in their television broadcasting systems.

Viewdata (Prestel)

The starting point in the BPO Viewdata development was a memorandum in 1970 from the then Managing Director, Mr. Edward Fennessy,

Figure 18.6. BT "Prestel" Viewdata equipment.

to Heads of Departments seeking means for making more effective, and profitable, use of the telephone system—noting that the customer's telephone and local network were used on average for less than 1% of the time. It was noted in the PO Research Department that, at the time, there were some 20 million telephones in use in the United Kingdom, and about the same number of television receivers, most of them in the same locations as the telephones. This immediately suggested the possibility of providing a visual information service by displaying in alphanumeric form on the television receiver screen information derived from a data bank at the exchange over the telephone line. The evolution of this idea into a working prototype Viewdata system was largely the work of Mr. Sam Fedida, supported by Mr. Keith Clarke and other colleagues in the PO Research Department. A laboratory "breadboard" model was first demonstrated at the Research Centre at Martlesham in 1971, followed by a full-scale demonstration in 1974 of a Viewdata system working over exchange telephone lines and providing access to a wide range of information sources.

The basic elements of a Viewdata system comprise:

- A television receiver with a Viewdata adaptor or decoder, either built in or added to an existing set
- A keypad, similar to a pocket computer, with the numerals 0 to 9, and * # symbols
- A data base and a computer for control of the system, located at a telephone exchange

The keypad gives the user interactive control of the system via a simple operating protocol, enabling the user to call up first a "menu" of the types of information available and then by successive stages of selection any one of 100,000 or more pages of information within seconds.

The control signals are sent to the exchange computer over the telephone line at a rate of 75 bits/sec; information signals are received from the exchange data bank at 1200 bits/sec. The decoder stores a frame (page) of information and displays it continuously in alphanumeric form on the television receiver screen. The display can make full use of the color capability of the television receiver. Recent developments have made possible the display of detailed graphics and photographic material.

The range of information sources that can be accessed are virtually unlimited and include remote data banks in addition to those at the local exchange. Typical sources are:

- Telephone directories and yellow pages
- Commodity prices
- For sale
- Travel timetables
- Leisure activities
- What's on
- News

The services that can be provided include:

- Banking and cash transfers
- Shopping from home
- Message (telegram) transmission
- Booking facilities

Figure 18.7. Prestel can provide updated "yellow pages" information electronically at the touch of a button.

Following the initial demonstration in 1974, it rapidly became apparent that Viewdata was a development of major importance, and a new PO Headquarters Department with Dr. Alex Reid as Director was set up to coordinate development and evolve business and marketing strategies. An important part of this work was to create a wide and useful data base by attracting a large number of information suppliers offering material of interest and value to users.

The term *Prestel* was adopted to describe the commercial form of Viewdata—the latter term then being used generally to describe visual information services, including the broadcast version "teletext." With the handover from Post Office Telecommunications to British Telecom the further development of Prestel was continued, with an increasing range of information sources and services becoming available to users.[7]

From the viewpoint of the provider of information, Prestel, unlike the broadcast teletext services, offered the possibility of charging the customer for the information selected according to its value.

Prestel's virtually unlimited information capacity, fast access, and capability for meeting local as well as nationwide information needs were additional advantages. Its use of existing local and national telephone networks facilitated speedy and economical provision of the service. The achievement of international standardization through the International Telecommunication Union would further extend its scope and value.

In the words of Max Wilkinson of *The Financial Times,* a leading financial expert:

> Few people now doubt that electronic information systems of the type pioneered by Prestel will sweep across the developed world, just as the printed word did after William Caxton's invention.

For its pioneering work on Prestel, British Telecom was presented with the MacRobert Gold Medal, Britain's premium engineering award, by HRH the Duke of Edinburgh in 1979, and the MacRobert Award of £25,000 was presented to Sam Fedida as the inventor of Viewdata (Prestel).

As to the future of Prestel, which functions in the analogue telephone network, there are now possibilities for providing similar services via the videophone and teleconferencing facilities of the ISDN.

Teletext (Ceefax and Oracle)

In the United Kingdom the Research and Engineering Departments of the BBC and IBA were quick to realize the possibility of using the vertical (interframe) interval of the television signal waveform for transmitting alphanumeric information that could be presented on the screens of domestic television receivers equipped with a decoder. (This interval had earlier been used to transmit specialized test signals such as the sine-squared pulse "K" rating signals pioneered by Dr. N. J. Lewis of the BPO Research Department in the 1950s and used for assessing the transmission quality of television links.)

The initial development of Ceefax and Oracle by the BBC and IBA proceeded independently of one another and of Prestel; however, after initial ideas had been clarified, it became apparent that compatibility between Ceefax and Oracle was essential, and a degree of compatibility with Prestel was desirable to gain the support of the television receiver industry and wide public acceptability.*

IBA's development of Oracle (an acronym for *o*ptional *r*eception of *a*nnouncements by *c*oded *l*ine *e*lectronics) was carried out in the Authority's Experimental and Development Department under the direction of the late Howard Steele, where the work was pioneered by Mr. Peter Hutt and supported by Mr. George McKenzie and other colleagues. A paper presented by Peter Hutt at an International Broadcasting Conference in London, September, 1972 (IEE Conference Publication No. 88) on the labeling of program sources in the complex IBA network by data inserted in the television waveform included, among other possible applications, the following: "Transmission of captions to special domestic receivers. Regional news and/or weather service distribution."

From this beginning, development work proceeded rapidly and IBA began its first "on the air" demonstrations of Oracle in April, 1973.

The BBC's development of Ceefax was carried out at the Kingswood Warren Research Laboratory, where Mr. S. M. Ewardson and J. P. Chambers made notable contributions. A simulated demonstration of the possibilities of Ceefax was given on the television program *Tomorrow's World* in October, 1972.

Compatibility between the Ceefax and Oracle systems was eventually reached through a joint BBC, IBA, and TV Industry Committee under the auspices of the British Radio Equipment Manufacturer's Association, and regular transmissions by the BBC and IBA commenced in the mid-1970s. Similar systems began to be developed in France, Germany, the United States, and Japan, and discussions took place in the Videotext Committee of the CCITT with a view to an international standard system. This was launched in 1976, based largely on the British Ceefax/Oracle systems, and designated the World System Teletext (WST) Standard.[8]

The broadcast teletext systems offer up to 200 or 300 pages of information (limited by the time available in the vertical interval of the television signal waveform). The desired page is chosen from an index

---

*The writer is indebted to Mr. Pat Hawker for an account of the early history of Oracle/Ceefax development.

page and accessed from a push-button unit by dialing a three-digit page number, with an average response time of a few seconds.

The menu offered by teletext within the scope of the 200 or 300 available pages is very wide, including weather and travel information, news, financial information and stock exchange prices, details of radio and television programs, what's playing in theaters and cinemas, holiday offers, and "for sale" notices.

Of particular value for the deaf are the captions that can be inserted when viewing some television programs. The teletext service itself is free to viewers and available throughout the area covered by normal television broadcast services.

It is fitting that the BBC and the IBA were in 1983 jointly presented with the Queen's Award for Technology for their pioneering work in creating the world's first successful teletext service.

REFERENCES

1. Daniel M. Costigan, *Electronic Delivery of Documents and Graphics* (Van Nostrand Reinhold, Princeton, NJ, 1978).
2. Ives, Horton, *et al.*, "Transmission of Pictures over Telephone Lines," *Bell Syst. Tech. J.*, April 1925.
3. *A History of Engineering and Science in the Bell System: Transmission Technology (1925–1975)*, edited by E. F. O'Neill (Bell Laboratories, 1985), Chapter 22.
4. "Post Office Launches Confravision," *Post Off. Electr. Eng. J., 64,* January 1972.
5. "Confravision by Satellite," *Post Off. Electr. Eng. J., 72,* January 1980.
6. "International Video Conference Service," *Br. Telecom Eng., 3,* October 1984.
7. *Prestel 1980*, edited by Dr. A. Reid (British Telecom, 1980).
8. G. O. Crowther, "U.K. Teletext—Its Evolution and Relation to Other Standards," IEE International Conference "The History of Television," Conference Publication No. 271, November 1986.
9. "The Codec Story," British Telecom, *Competitive Edge*, Issue 4, May 1991.

19

Information Technology and Services

Data Communication

To an increasing extent, social, economic, and political activities in the modern world are becoming dependent on information technology (IT)—the means by which information in the form of data is stored, processed, and accessed. The advances in telecommunications outlined in earlier chapters have removed the barrier of distance and reduced the costs of transmission so that users and data banks can be remote from one another. The microchip has enabled vast amounts of data to be stored, rapidly processed, and accessed at costs that continue to fall. And in the future IT may well have a powerful influence on where people live and work, and on the environment itself by minimizing the need to travel to communicate (Refs. 1, 2, and Chapter 21).

IMPACT OF TELECOMMUNICATIONS AND INFORMATION TECHNOLOGY ON THE WORLD SCENE

The advances in telecommunications techniques and IT made in the last two or three decades have already had a dramatic and significant impact on the world scene. It is relevant to recall a few instances that have caught the imaginations of, and are stored in the memories of, millions of people:

- The telephone "hot line" between Presidents Kennedy and Khrushchev that staved off World War III in the Cuban missile crisis of 1962
- The television pictures by the TELSTAR satellite covering, in agonizing clarity, the assassination of President Kennedy in 1963
- The virtually instantaneous transmission by electronic means of vast sums of money that facilitated the overrapid computer-controlled buying and selling of shares that precipitated the world stock-market crash of 1987
- The live television pictures of the first moon landing and the conversation between President Nixon and astronauts Neil Armstrong and Buzz Aldrin in 1969
- The detailed pictures of the most distant planets of the solar system made possible by picture-storage, slow-scan IT techniques
- The live action pictures and commentaries by news correspondents, via satellites and compact microwave transmitting equipment, from war and disaster areas worldwide

And it may well be that the existence of intelligent life on remote planets of our galaxy will be revealed by the search of the radio spectrum now being conducted using large microwave aerials and IT-based signal recognition techniques. But perhaps one of the most exciting uses of IT has been to process the pictures of galaxies at the most distant edge of the universe, revealing what they were like at the time of the big bang 15 billion years ago.

However, it must be admitted that there is a danger related to the uses of IT—exemplified by the activities of "hackers" who use their knowledge of IT and system organization to gain illegal access to data banks. One such incident concerned the Duke of Edinburgh's personal correspondence data bank, but a much more dangerous case involved the "War Games" data bank of the U.S. military at the Pentagon! And even the private conversations of the U.K. "royals" involving mobile radio have not been spared.

TEAM DEVELOPMENT
OF INFORMATION TECHNOLOGY

The ongoing development of the telecommunications applications of IT has been largely based on the basic device technology, such as the microchip, and software/operating system techniques common to the computer field. The development has tended to be carried out by teamwork in the laboratories of large telecommunication organizations such as British Telecom and AT&T/Bell, and computer firms such as IBM, rather than by readily identifiable individuals, as in earlier chapters of this book. Furthermore the need for international standardization of technical parameters to facilitate the setting up of worldwide networks has meant that this aspect of development has to be carried out via international committees, for example of the International Telecommunication Union.

It is perhaps appropriate at this point to examine in more detail the nature of IT and the telecommunication services it can provide.

INFORMATION TECHNOLOGY AND
TELECOMMUNICATION SERVICES

IT is primarily concerned with the manipulation of data. For present purposes, "data" refers to spoken or written words, letters, punctuation marks, numbers and other symbols, graphic and photographic material, and video signals, in analogue or digital form.

"Information" is an assembly of knowledge, facts, ideas, instructions, and the like which can be expressed by data, and which is meaningful to, but not necessarily new to, a recipient. It will be recalled from Chapter 13 that Claude Shannon defined the basic unit of information as the "bit," corresponding to an on–off or yes–no signal.

The elements of an IT system comprise:

- Computers, in which information is stored as data and processed as determined by a control program

- Telecommunication networks, with terminal equipment for the insertion, selection, and display of data, data transmission and switching systems

An important feature of developments during the last decade or so has been the convergence of computer and telecommunication technology, exemplified by the common use of microprocessors, stored-program control, and digital methods. The visual display units (VDUs) that are a feature of computer systems owe much to the development of the television picture tube, while the computer printer is related to the facsimile machine. Thus, it might be said that the innovators and pioneers of key advances in telecommunications have benefited from the development of computers, and vice versa. And both groups have contributed to the creation of IT.

The range of telecommunication services available under the generic category of "information technology" is wide and expanding; they include the following.

1. Audio Conferencing

Audio conferencing facilities enable three or more ordinary telephone users at different locations to be linked on a dial-up basis via the public switched telephone network or via private circuits. A more extensive audio conferencing facility enables larger goups, e.g., in halls, to confer via loudspeakers, microphones, and voice-switching equipment. The ISDN now under development enables audio conferencing to be provided more quickly, with greater economy and better speech quality than is possible over 3-kHz-bandwidth telephone channels; advanced codec design for analogue speech signal to digital conversion enables high-fidelity speech transmission to be achieved over 64-kbit/sec digital channels.

2. Teleprinter Transmission (Telex)

This provides an automatic electric typewriter public switched service with transmission by digital signals using internationally agreed (CCITT) codes and alphanumeric alphabets, typically at a speed of 50 bits/sec.

The service, initially based on relatively slow electromechanical machines, is gradually being superseded by faster, quieter electronic systems using facsimile and visual display (VDU) techniques.

3. Data Transmission

Data transmission at speeds of 9600 bits/sec or more can be made over the voice circuits of the public switched telephone network, via modems that convert the digital signals to keyed audio tones. Higher speeds, typically 64 kbits/sec or multiples of this up to 2 Mbit/sec, are available over the ISDN, the faster speeds being valuable for the transfer of large-capacity data banks and the provision of high-definition facsimile, e.g., for whole-page newspaper transmission.

4. Viewdata (Prestel)

The British Telecom "Prestel" system provides domestic and business users with access to visually displayed information from a wide range of data banks having virtually unlimited information capacity, over the public switched telephone network. It is a fully "interactive" system giving the user complete control over the information selected, which can include graphic and still-picture displays as well as alphanumeric information (Chapter 18).

5. Telewriter and Electronic "Blackboard" Systems

The telewriter is a narrow-bandwidth document transmission system that enables handwriting or sketches to be transmitted over the public switched telephone network or private circuits, and displayed simultaneously as written—on a sheet of paper or on a large screen by overhead projection if required. It uses a pen moving over a wire mesh writing pad that translates the pen position into electrical X, Y coordinates, which can then be used to control the position of a writing pen at the receiving end.

A version of the telewriter principle can be used to create an electronic "blackboard," e.g., for teaching purposes. The narrow-bandwidth visual signal is transmitted together with voice over a telephone circuit; it has also been used via sound radio broadcasting channels for teaching purposes in developing countries.

6. Facsimile Transmission

Facsimile systems can provide good-quality transmission of hand-written or typescript material and diagrams over telephone circuits, with a transmission time of less than one minute for an A4 sheet. Faster transmission and higher quality, including the transmission of halftone photographic material, are possible using more sophisticated coding techniques over ISDN. Documents can be sent to several destinations simultaneously, and delayed (e.g., overnight) transmission is possible to take advantage of cheaper rates (Chapter 18).

7. Television Conferencing

As described in Chapter 18, group-to-group television conferencing is available over intercity and international cable, microwave, and satellite television links and will become increasingly economical as more advanced coding techniques reduce the video bandwidth requirements. The use of optical fiber systems, with their wide available bandwidth, will further reduce transmission costs. The ISDN will provide person-to-person "videophone" conferencing, as well as group-to-group higher-definition video conferencing (Chapter 18).

8. Mobile (Cellular) Radio and "Phone-Point" Systems

The last decade has seen a massive development in the United Kingdom, the United States, Europe, and Japan of mobile radiotelephone systems using hand-held portable equipment operating at frequencies in the UHF band via an extensive overlapping "cellular" network of land-based stations (see Chapter 20).

These provide both-way access to the national telephone systems from cars and trains using standards and procedures set by the International Telecommunication Union. Communication satellites may offer better coverage than land-based radio stations, but with some restriction imposed by the available radio spectrum on the scope for exploitation.

A further development—"Phone-Point"—enables calls to be made from hand-portable short-range radiotelephones via a number of fixed points in busy city centers, e.g., railway stations.

9. "Electronic Mail" and Information Networks

The 1990s has seen a remarkable worldwide development of "electronic mail" (e-mail)—a person-to-person immediate communication technique via keyboard and visual display/paper copy. E-mail is character based, although digitally encoded images may also be transmitted. It uses alphanumeric presentation, with digital transmission using the ASCII code (see below) over conventional telephone circuits. As such it has features in common with the earlier "telegram" and "telex" (teleprinter) services but provides much speedier and more flexible service. For instance, e-mail offers convenient two-way communication between "mailboxes." It also offers a "bulletin" or notice board containing information of permanent or semipermanent nature, relevant to a given organization or group. And above all it is a relatively low-cost service, since it can be provided over ordinary telephone circuits.

However, it must be emphasized that other people have access to your computer and messages are thus open to interception for purposes that may not be legitimate. Furthermore the connection of your computer into an unlimited network may make it vulnerable to the injection of a computer "virus," i.e., illegal operating instructions that destroy or garble information.

The success of e-mail is in part related to the setting up of national and international information networks under such titles as "Demon" and "CompuServe" which provide computer-controlled access via the worldwide "Internet" to a wide variety of data bases and user groups, generally over existing national and international telecommunication facilities.

THE CONVERGENCE OF INFORMATION AND COMPUTING TECHNOLOGY

The development of telecommunication-based information processing and data-bank accessing systems such as those outlined above has been made possible by the parallel development of computing technology. The latter includes the following basic and highly significant developments:

- The "ASCII code" (American Standard Code for Information Exchange), in which the lower- and uppercase letters and the numerals, the punctuation marks, and spacing between letters are each allocated a number up to 100 which also identifies a key on the computer keyboard

- The "compact disc" (CD) which enables many millions of bits of information to be recorded on spiral tracks on a disc a few centimeters wide and read either magnetically or by a fine laser beam; the computer operating system can be arranged precisely to locate and record, or read, on the disc selected areas of information—the system is known as "CD-ROM" or read-only memory
- The general concept of computer control by "software," the digitally encoded instructions, recorded on disc or tape, which instruct the computer to carry out prescribed functions and operations
- The specific software concept known as "Microsoft DOS," where DOS refers to Disc Operating System and Microsoft is a U.S. company trademark (the origin of Microsoft DOS, which is used for computing worldwide, has become an essential element in IT, and has resulted in a billion-dollar industry in the United States, is described below)
- Linked with "Microsoft" is the important "windows" concept which enables chosen symbols (icons), alphanumeric information, still or video pictures to be placed as required in selected areas of a VDU screen; a screen cursor, controlled by a manually operated control or "mouse," enables an icon denoting and activating a chosen function to be selected at will; these concepts are particularly important for the "multimedia" development of information services
- The advent of the low-cost personal computer/word processor made possible by the falling costs of microchips, which has not only replaced the large and costly "mainframe" computer for many purposes but also has enabled IT to reach into the home as well as the office; in the United Kingdom Alan M. Sugar and in the United States Ed Roberts have made valuable contributions to this development, the former with his Amstrad PC/WP

The Origins of "Microsoft DOS"

Much of the credit for the creation of Microsoft DOS must go to the vision and enterprise of young Bill Gates who, with college friend

Paul Allen, realized in 1975 that the then existing commercially available computers lacked the flexibility to handle personal computing/word processing. From Seattle, they produced the first microcomputer software that helped to launch the PC revolution.[8]

By 1979 Gates and Allen had licensed their software to other U.S. computer manufacturers; rather late in the day the giant computer firm IBM realized that its future in the PC field lay in the hands of Gates and Allen and a small company called Microsoft holding basic patents in PC software.

From these beginnings, and with a computer operating system called QDOS bought from another Seattle company and translated into Microsoft by Gates and Allen, in 1981 IBM launched MS-DOS, a PC that became familiar to computer users all over the world. Other U.S. computer firms entered the battle, and Apple developed a PC with an improved operating system offering "icons" and "windows."

By progressive innovation and enhancement of its original patents, Microsoft has maintained a dominant position in the computer software field—a position that, however, is currently (1994) being challenged by others via the U.S. Department of Trade and Industry and the U.S. Justice Department as possibly harmful to competition in the computer industry.

THE INTEGRATION OF INFORMATION TECHNOLOGY AND HOME/BUSINESS COMMUNICATION SERVICES

1. Home Services

As the range of information and entertainment services available to customers in their homes expanded, it became logical to seek ways of providing these on a common local distribution network since this could reduce the costs of provision and be more convenient to the users, as compared with separate provision of the various services.

One such forward-looking study was conducted by the "Subscribers Apparatus" Group at the BPO Research Department, Dollis Hill, in the late 1960s and a prototype laboratory demonstration model was made. This embodied a broadband coaxial cable ring main providing a digital highway and to which customers' premises were linked in conveniently located groups over feeder cables or short-range microwave links, via traffic concentrators.

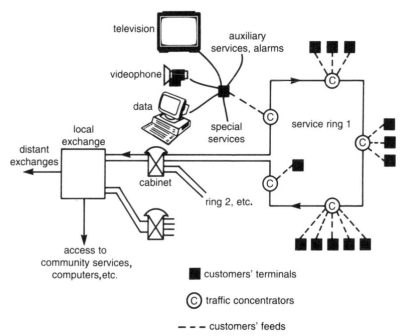

Figure 19.1. Prototype broadband local distribution network using a digital coaxial cable ring main (British Post Office Research Department, 1968–1972).

The services then envisaged included videophones, data VDUs and access to community data banks and computers, and broadcast television and auxiliary services such as fire alarms. The project, which was recorded in a color film "Telecommunications 1990," was perhaps ahead of its time. Nevertheless, the advent of the microchip, optical fiber cables, and the spread of digital transmission and switching throughout the trunk network have now made it a practical reality.

To these entertainment services might be added on-demand access to the "video library" service discussed later in this chapter.

Current studies of "the home of the future" envisage a wide range of functions that IT could play, ranging from voice and visual communication, information access, entertainment services, shopping and banking from home, home security alarms, and "working from home."

However, at the present time political ideology and the claimed merits of competition appear to inhibit in the U.K. the full integration of entertainment and communication services in the home.

2. Business Services

The impact of information and computer technology in the business field, in banking, and in government at national and regional levels is substantial and expanding. In the business field it provides immediate communication between individuals and groups, rapid access to data files on such diverse topics as sales, stockpiles, cash flow, personnel, group performance, and stock-exchange share prices, and provides a mechanism for large and small cash transactions and transfers. Within a business it can provide a range of services including directory inquiries,

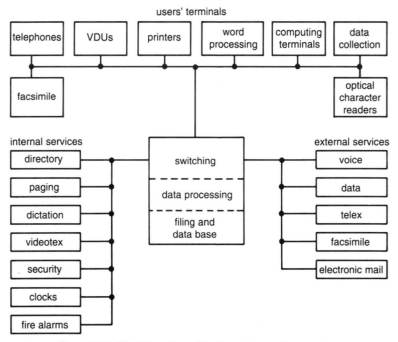

Figure 19.2. The integration of business information services.

paging, dictation and videotex (display of information on VDUs), and security and alarm facilities. By voice, facsimile, telex, electronic mail, and audio/video conferencing services, these facilities can be extended to other locations, e.g., from business headquarters to other offices, factories, and sales units, without regard to distance.

One commentator has noted:

> the power of information technology to supply managers with accurate information on the day-to-day performance of groups . . . and the ability of information technology to condense information into readily understandable charts and figures . . . allows chief executives to control larger teams than was possible using pre-computer management methods.

However, there is a note of caution in the familiar phrase from the computer field—"garbage in, garbage out!" i.e., the output from an information system can be no more reliable than the data supplied to it.

DATA COMMUNICATION

The theory of data communication involves a logical structure that enables data terminals (data sources/receivers) to be interconnected via transmission networks involving one or more computers or switching centers. The general objective is to ensure that message information—data—reaches its intended destination(s) without error and with a minimum time occupancy of the transmission network. As the numbers of terminals and computers increase, so does the complexity of network organization if the objective of efficient use of transmission paths is to be maintained. The reader is referred to specialized texts, e.g., via Refs. 3 and 4, for a detailed account; for present purposes the following modes of operation of data networks may be distinguished.

1. Circuit Switching

This is the mode generally used in telephone systems where calls are "dialed up" and switched at one or more exchanges via electromechanical or electronic switches, the transmission path being held "open" for the duration of the call, irrespective of whether message information is passing or not. From the latter point of view it is not a highly efficient use of the network; however, the rate of data transmission is limited only by the bandwidth of the transmission path and, since no "store and forward" operation is involved, there is no traffic congestion.

2. Message Switching

Message switching is particularly adapted to multiple-node networks, i.e., involving several computers and transmission paths. Each computer node has processing capacity that enables it to:

- Interpret an address at the head of each message
- Temporarily store each message
- Send the complete message to the computer attached to a selected node or send it on to another node

This mode of operation is akin to the "store and forward" technique used for many years in the telex system, with human operators providing the "processing." It may, however, suffer from message clogging if the buffer stores at nodes have inadequate capacity to handle particularly long messages.

3. Packet Switching

This seeks to achieve a highly efficient use of the network by splitting up messages into short packets of data, each with a header containing information about the identity of the message, the sender and receiver, the sequence of the packets in a given message, the message priority, and the identification of the last packet of a message. Clearly, this involves additional data but the overall benefit of avoiding congestion from long messages represents a worthwhile advantage.

THE INTEGRATED SERVICES DIGITAL NETWORK

The economics of data communication and the uses of IT are being profoundly influenced by the development of ISDN worldwide.

ISDN provides a wide and growing range of "dial-up" modes of communication, including telephony, facsimile, videotex, audio and television conferencing (Chapter 18), and data-base transfer. Thus, ISDN is essentially multipurpose and is also capable of multipoint operation, e.g., for "broadcast" distribution of messages. The structure is based on the 64-kbit/sec bit rate for PCM digital speech transmission, which can be used in integral multiples of bit rate for wider bandwidth services, e.g., group-to-group video conferencing. Furthermore, since charges can be related to duration of use and bit rate, ISDN can provide an economical service for the user.[7]

Milton Keynes New Town and Bishop's Stortford: Test-Beds for ISDN

Milton Keynes New Town, Buckinghamshire, England, has been described as "a town of the future, a showcase for new technology . . . one of the cornerstones of its success has been the forward thinking that has gone into the town's telecommunication and information technology."[5]

By early 1988 Milton Keynes became the first urban area in the United Kingdom to be served entirely by a digital network with all of the benefits of speedier connections and short-code dialing, better voice quality, and faster data transmission. The services available include teleconferencing and digital data services such as:

- "KiloStream" for electronic mail, credit verification, fast facsimile, and computer links to remote terminals
- "MegaStream" for high-speed data transmission at rates up to 2 Mbit/sec, e.g., for computer-to-computer data transfer, interconnection of digital private branch exchanges
- "Packet Switch Stream," the national packet-switched network for nonvoice communication at distance-independent tariffs, which is linked to similar networks in more than 50 other countries
- "SatStream," which provides leased two-way circuits at rates between 56 Kbit/sec and 2 Mbit/sec to Europe and North America via small satellite dishes

The Milton Keynes cable network also provides sound and television services from the BBC and ITV, and satellite television channels for the home, together with a locally produced information service using "photovideotex" for the presentation of local news, shopping, housing, job and leisure information, and advertisements. Besides being available in the home, this service is presented on public viewdata terminals sited around the town.

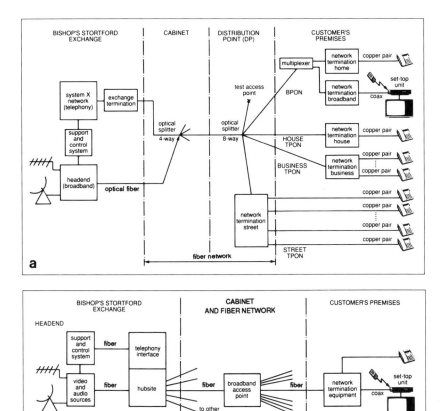

Figure 19.3. Bishop's Stortford integrated services field trial on optical fiber cable (1990). (a) Passive optical fiber local network telephony (TPON) and later broadband (BPON). (b) Active optical fiber star-switched broadband local network.

The town of Bishop's Stortford in Hertfordshire is also the setting for a comprehensive field trial involving the provision of integrated services to domestic and business users. Two approaches are being explored, with the aim of identifying the most economical method for meeting various customer requirements.

One approach uses a "passive" optical fiber local distribution network, i.e., one not involving active devices, to provide telephony and low-bit-rate services at minimum cost while permitting upgrading to broadband services such as television in the future. Copper wire pairs or coaxial cable are used for the relatively short connections at the customers' premises.

Another approach uses an active switched-star network, similar to that in the Westminster cable-television trial (see below), adapted to include telephony; this method uses optical fiber right up to the customers' premises.

THE ON-DEMAND INTERACTIVE VIDEO
LIBRARY SERVICE

In spite of a growing multiplicity of continuously running television channels by cable and satellite—each heavily loaded with material of only marginal interest to many viewers—there is an increasing public interest in the use of bought or rented videocassettes offering full-length recordings of plays, films, documentary and other material. These allow the viewer to choose both what to view and when to do so—free from advertisements. This suggests that an on-demand interactive video library service, provided electronically over cable or a local area microwave radio network, could offer an immediately available, convenient to use, and wider choice of viewing material. Payment for the service could be closely linked to the duration and quality of the material viewed, for example with higher rates for first-run films already acclaimed by the critics.

In 1986 British Telecom initiated a pioneering field trial of such a service over cable in the Westminster area of London, mainly to hotels. The viewer was presented on the TV receiver screen with a "menu" of titles of items to view from which he or she made a choice by operation of a push-button keyboard. The video material offered was recorded on Philips "LaserVision" discs, selected exclusively for each customer by mechanical selection at the video library center on a machine reminiscent of the jukebox used to play audio records. The system was designed initially to accommodate up to 250 simultaneous sessions from some 10,000 customers, over the BT switched-star TV cable network.[6] The switched-star network concept itself represents an important contribution to the economical distribution of television by cable; for his work on the system Bill Ritchie of the BT Martlesham Research Laboratories was awarded the Martlesham Medal in 1988.

Clearly, the future of the video library concept will depend on a much more sophisticated use of electronic techniques to store and access video material—here is an important new goal for innovators which could lead to a new, and ultimately more satisfying, mode of viewing recorded video and television programs.

British Telecom Trials of "Video on Demand" over Customers' Telephone Copper Pairs

Far-reaching developments, based on advanced video digital coding techniques pioneered, among others, by AT&T in the United States and British Telecom at its Martlesham, Suffolk Laboratories, are enabling TV broadcast-quality video signals to be transmitted over ordinary copper pairs used to provide telephone service to customers' premises. The digital compression achieved enables the bit rate to be reduced from the 100 Mbit/sec hitherto needed for good-quality video to some 1.5 or 2 Mbit/sec. The technological consequences of this advance are twofold:

- The digital signals can be transmitted over several kilometers of telephone copper pairs without significant impairment of picture quality.
- A video of an hour or more can be stored on compact discs or microchips, with random access on demand; this gives the prospect of more than 1000 films being conveniently stored in memory for access by a single high-speed computer.

The video digital compression techniques are being studied with a view to worldwide standardization by the International Standards Organization (ISO) and the International Electrotechnical Commission (IEC); they have potential application not only to "video on demand" but also to TV broadcasting and other video services.

The BT field trial of "video on demand" at Kesgrave, Suffolk, at present (1994) offers a modest choice of video material, but is capable of expansion as specialized ROM microchips become available. Equally important for the future are the improvements in real-time encoding of video signals, leading to picture quality improvement in low-bit-rate modes, now being studied.

The proven 1.5- or 2-Mbit/sec capability of the copper pair network clearly has potential for fast data, as well as video, services. These are not necessarily in competition with the wideband services available on optical fiber networks, but can serve as a valuable bridge to the future.

REFERENCES

1. Murray Laver, *Computers, Communications and Society* (Oxford University Press, London, 1975).
2. Murray Laver, *Information Technology: Agent of Change* (Cambridge University Press, London, 1989).
3. C. P. Clare, *A Guide to Data Communications* (Castle House Publications Ltd., Tunbridge Wells, Kent, 1984).
4. "Information Networks," *Br. Telecom Eng. J.*, 9, Pt. 1, April 1990.
5. "Building Tomorrow on Firm Foundations (Milton Keynes)," *British Telecom World*, September 1989.
6. "On-Demand Interactive Video Services on the British Telecom Switched Star Cable-TV Network," *Br. Telecom Eng. J.*, 4, No. 4, October 1986.
7. J. M. Griffiths, *ISDN Explained: World-Wide Network and Application Technology* (Wiley, New York, 1990).
8. This account of the origin of Microsoft DOS is based on a discussion and interview with Bill Gates, Chief Executive of Microsoft, broadcast by BBC in its "In Business" program, June 1994.

20

The "Mobile Radio" Telecommunicators

EARLY HISTORY OF MOBILE RADIO COMMUNICATION

Mobile radio communication may be described as communication with and between individuals who may be in land vehicles, ships, trains, or aircraft, via telegraph, telephone, facsimile, e-mail, or other services.

The history of mobile radio communication is as old as that of radio itself—much of Marconi's work in the early 1900s was stimulated by the need, especially for safety reasons, for communication with and between ships, a need that radio alone could satisfy at that time (Chapter 7). From the 1920s onwards came the worldwide development of coast radio stations for communication with ships, using the MF and HF regions of the radio spectrum and amplitude modulation (AM).

World War II gave a strong impetus to the development of VHF radio technology (30–300 MHz), both for communication in the fighting services and for radar. The 1950s saw the widespread use of VHF for civilian land-mobile radio services such as the ambulance, police, and fire services, and commercial users. Typically, these VHF systems used narrow-deviation frequency modulation with 100-watt base stations and provided ranges up to some 50 miles. The invention of the transistor in the 1950s, and later the microchip, enabled the size and power requirements of the equipment to be substantially reduced and the reliability improved.

Similar developments took place in maritime mobile radio, e.g., for shipborne use, with international standardization of the technical parameters and operational procedures via the International Telecommunication Union (ITU) and the International Radio Maritime Committee (CIRM). However, the expansion of these services and the growing demands on the VHF radio spectrum, which included broadcasting as well as mobile radio, made it necessary to seek new approaches to mobile radio communication, e.g., using frequencies in the UHF (300–3000 MHz) range where more frequency space is available.

THE UHF "CELLULAR" LAND-MOBILE RADIO SYSTEM

However, a move to UHF brought its own problems. Shadow effects, e.g., of hills and tall buildings, are more pronounced than for VHF and the area effectively covered from each base station is thus restricted, generally to a radius of a few miles. Furthermore, because the wavelength is shorter, wave interference fading ("Rayleigh" fading) as seen from a moving vehicle occurs faster than for VHF.

A major step forward was taken in the early 1970s when the American Bell System submitted a report to the U.S. Federal Communications Commission proposing a "cellular" mobile radio system using a multiplicity of marginally overlapping base-station areas, each with a radius of a few miles. Except in the immediately adjacent base-station areas, this permitted reuse of frequencies in the various cells of the network and thus economy in spectrum utilization.[1]

The base stations were connected into the public switched telephone network via fixed cable or microwave links; because of the multiplicity of mobile users in the many cells of the network and the time dispersion of their calls, a limited number of "trunk" paths could provide a satisfactory grade of service. (The trunking traffic theory due to Erlang is relevant; see Chapter 14.)

An important technological development was the "frequency synthesizer," based on a combination of digital techniques and phase-locked loops, which enables operating frequencies to be changed quickly and automatically as the mobile moves from cell to cell.[2]

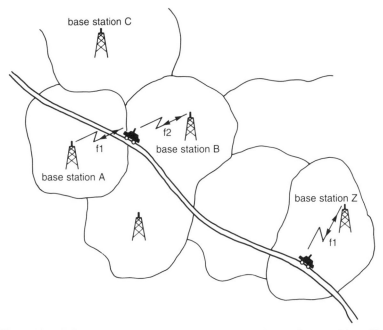

Figure 20.1. Cellular mobile UHF radio system. Frequencies can be reused in sufficiently widely spaced cells such as f1 and A and Z, but not in adjacent cells such as
A and B.

As in other communication systems, cellular mobile radio has
moved on from analogue (frequency modulation) to digital techniques
offering important advantages, notably:

- Low-bit-rate encoding with forward error correction which minimizes the impairment of speech quality caused by fading
- Encryption to ensure speech privacy

International standardization of the technical parameters [e.g., via
the European Posts and Telecommunications Conference (CEPT) "Global
System for Mobile" (GSM) group] is required if mobile users are to
move freely from country to country.

Looking to the future, there are interesting possibilities for using
millimeter microwaves, e.g., in the absorption band around 60 GHz, to
provide more capacity as demand grows. Although the cell size is reduced, the mobile system might be integrated with the use of millime-

ter microwaves for providing local area fixed networks to customers' premises.

SATELLITES AND MOBILE RADIO COMMUNICATION

For more than 15 years the international maritime satellite organization IMARSAT has been providing telephone communication with ships and aircraft. It is now (1994) poised to provide worldwide telephone/ fax/data communication with land-mobile units, as well as ships and aircraft, using compact "briefcase"-size mobile equipment, or even "hand-held" equipment similar to that used in terrestrial cellular radio communication systems.

Two types of satellite system are under consideration:

- 4 to 6 large satellites in an equatorial, geostationary orbit at a height of 36,000 km
- 12 smaller satellites at a height of 10,000 km in two orbital planes inclined at 47° to a third, polar plane.

A number of other orbital configurations are under consideration by various U.S. companies, using digital modulation techniques.[3]

Such satellite systems could provide virtually worldwide coverage, including land areas not at present covered by terrestrial cellular systems or which are unlikely to be so covered. However, the large area covered by each satellite—which would include far more mobile users than each cell of a terrestrial cellular system—could give rise to problems of frequency allocation for large-scale utilization.

REFERENCES

1. Bell Laboratories Report to the Federal Communications Commission, "High Capacity Mobile System Technical Report," 1971. Also *Bell Syst. Tech. J.*, January 1979.
2. "Frequency Synthesisers—A Review of Techniques," *IEEE Tech. Trans., COM-17*, 1969, pp. 357–371.
3. D. B. Crosbie, "The New Space Race—Satellite Communications," *IEE Rev.*, May 1993.

21

Telecommunicators
and the Future*

This review of innovators and innovation in telecommunications and broadcasting systems has shown their evolution from the primitive beginnings of electrical science in the 18th century to the comprehensive, powerful, and worldwide communication systems of the late 20th century, which are now indispensable to the social and business worlds and the processes of government.[1,2]

The technology and system concepts that have been created would appear to provide well for existing communication service needs, both in types of service and by scale, and even to anticipate in some degree future needs. Is there scope for innovation in the future?

Telecommunications in the last 50 years has been mainly "technology led," i.e., the services provided were stimulated by the invention of new devices such as the microchip, coaxial cable, microwave radio-

---

*Some of the material for this chapter was stimulated by the forward-looking "Innovation '94" symposium presented by British Telecom at their Martlesham Heath Laboratories in May, 1994.[3]

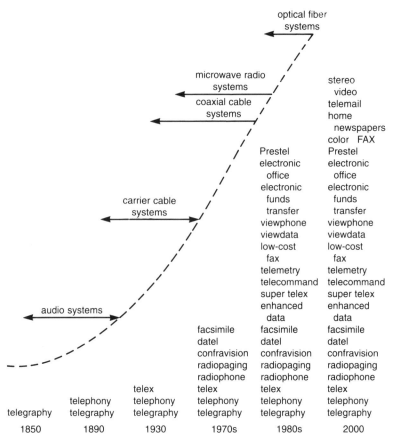

Figure 21.1. The growth of telecommunication services and the matching transmission systems from 1850 to the year 2000.

relay and satellite systems, and computer-controlled electronic switching exchanges. However, there are ongoing developments in micro- and optotechnology that may well stimulate the provision of new services, as well as reduce the costs of providing existing services.

The history of the development of the microchip in the last decade has demonstrated a dramatic and continuing increase in the number of transistor elements that can be accommodated on a single 1-cm square of silicon—doubling each year or so—and a substantial reduction in the cost of providing a given memory or switching/processing function. In the early 1970s the number of such elements on a chip was about 2000 with a line spacing of 6.5 μm and typical memory chips held 1024 bits of information with 8-bit processing. Today (1994) 35 million transistors per chip with a line spacing of 0.5 μm and 64-bit processing are state-of-the-art, with greatly increased operating speeds. And by the end of the century 250 million-bit microchips are predicted. Such increases in memory and processing power, together with improved video coding methods, will, for example, enable high-quality video programs of an hour or more to be stored on a single microchip— opening the door to economically viable "on demand" video viewing. The microchip technology initiated by Robert Noyce of Intel and Jack Kilby of Texas Instruments in 1959 (Chapter 12) has indeed borne fruit!

Nevertheless, development in the next 50 years will be mainly "software led"; this will involve increasingly sophisticated computer-type programs for the control of transmission, switching, and network organization to facilitate the growth of existing services and provide new services tailored to meet specific user requirements. The trend is thus toward an "intelligent network," capable of responding on demand and cost-effectively to a wide range of user requirements.

But we are now at the beginning of an "information revolution," christened "cyberspace" by the media, in which digital highways on satellites and optical fiber cables link computer-controlled switching nodes to create a worldwide network—the "Internet"—with unlimited access to information data banks and direct person-to-person audio, facsimile, text, or visual communication. In the words of John May of the *Sunday Telegraph*,[4]

> We are moving from a world where events are unrelated, and where hierarchical power structures predominate, into a world where the principal organizing structure is a network. In a network, everything—people, ideas and events—is linked, connective and inter-active. This fundamental change is the most significant aspect of what we are experiencing and will have the profoundest effect of all on our development as a species.

The worldwide Internet is discussed later in this chapter, as are the implications of the advance of information technology on the future of civilization.

MULTIMEDIA SERVICES: THE INTEGRATION OF TELECOMMUNICATIONS AND BROADCASTING

The evolution in recent years of integrated service digital transmission and switching systems (ISDN), with an internationally standardized digital hierarchy accommodating a wide range of bit rates and services, provides a flexible framework within which telephone, facsimile , e-mail, audio and visual teleconferencing, and video services—conveniently referred to as "multimedia" services—can be provided on a common transmission path. Equally the ISDN could accommodate the sound and television services traditionally provided by radio broadcasting, and the forward-looking "view on demand" television service that could eventually replace continuously running television programs. The digital basis of the ISDN lends itself to advanced video coding techniques that reduce bit-rate requirements for defined standards of picture quality, facilitate video signal storage on microchips, and shorten facsimile signal transmission time.

While even the existing telephone copper pairs to customers' premises offer potential for good-quality video transmission and some scope for multimedia services, the optical fiber cable with its wide-frequency bandwidth has virtually unlimited scope for such services. Millimeter microwave radio and short-range free-space optical links both offer potential, as yet unexploited, for local area distribution networks.

The case for integration of telecommunication and broadcasting services in common distribution networks, both at a local level to customers' premises and at an intercity level, would seem undeniable in terms of economical use of resources, cost, and convenience to users. The barriers to integration appear to be political/ideological rather than economic/technical.

Television (Video) Conferencing

The person-to-person and group-to-group television conferencing facilities described in Chapter 18 mark the beginning of a visual service with substantial potential for development and growth. In particular may be noted the scope for innovation in large-screen three-dimensional presentation with stereophonic sound for group-to-group conferencing to create a very real sense of "you are here." This could be further enhanced by development of a technique to avoid the lack of "eye-to-eye" contact that occurs with a single camera.

The On-Demand Video Library Service

The on-demand video library service—in which the viewer has an immediate access over cable or local microwave radio to a wide range of recorded video programs and video information sources in one or more central libraries—could provide what many would regard as a more acceptable form of video viewing than can be offered by a multiplicity of conventional, continuously running broadcast television channels; it could make available, for example, high-quality first-release films on a pay-for-what-you-view on-demand basis, free from advertising material.

But, as noted in Chapter 19, this service does not wait for the provision of cable or optical local networks. A major step forward has been made (1994) with the development of sophisticated digital coding techniques enabling good-quality video transmission to be made on telephone copper pairs to customers' premises. In view of the universal and immediate availability of this type of local distribution network, the prospects for the future are, as noted above, highly promising for other integrated digital services as well as on-demand video.

Domestic Television Viewing

Innovation in broadcast television domestic viewing is likely to continue the trend to higher-definition and larger, flat-screen displays, making use of improved coding and redundancy removal to minimize transmission bandwidth requirements, using cable and satellite channels.

A virtually unexplored television viewing device is the video analogue of the personal headphone audio listening device, in the form of miniature screens built into the viewer's spectacles. Accompanying headphones could provide high-quality stereo sound. The possibilities are interesting—such a device could offer the illusion of being in an unrestricted three-dimensional world, entirely cut off from the viewer's immediate surroundings.

Virtual Reality

The personal video viewer device leads to the concept of "virtual reality" in which a viewer, equipped with video headset and hand-manipulated sensors, is free to explore a picture data bank at will. The data bank would contain the vast volume of picture information needed, for example, to view from all angles a building or for a scientist to examine a DNA molecular model.

However, specialized applications of virtual reality viewing are beginning to appear, for example in the remote viewing at will from headquarters of work on a building construction site, or providing from a hospital of skilled guidance for a remote surgical operation.

The Electronic Newspaper

The electronic newspaper of the future will probably use a flat display unit based on liquid-crystal and microchip techniques, thin and light enough to be comfortably hand-held, and linked by short-range radio to the telecommunication network for its sources of information. The displayed information could be transmitted overnight for reasons of economy, and updated as required during the day. An associated facsimile machine could provide hard copy of selected articles. Billing could be itemized and charged in terms of usage and items accessed. Such a system could offer "newspapers" of virtually unlimited scope and variety, of local, national, or overseas origin. Clearly, a regulatory system would be desirable to avoid a monopoly situation.

The social and environmental benefits could in time be substantial. A major advantage over the present-day large-scale use of paper for newsprint would be a slowing down of the rate of destruction of the world's forests. The uneconomical use of manpower and the congestion of roads necessitated by today's cumbersome, slow, and inefficient newspaper distribution system would be replaced by a speedy, energy-efficient electronic system.

An End to Scientific Journals?

Already serious consideration is being given to the future of scientific journals in view of increasing production and distribution costs, the limited scale of the market, and the need for speedier dissemination of information in a fast-moving scientific world.[5]

However, there seems no reason in principle why an electronic system, e.g., one similar to that outlined above for the electronic newspaper or a Viewdata/Prestel system, should not be adapted for the preparation and distribution of scientific and similar specialized journals.

The advantages of doing so could be appreciable; not only could the costs and time involved in distribution be reduced, the editorial preparation of journals, including peer review and acceptance, could be speeded up. The immediate availability of hard copies of selected articles and data would be welcomed by scientific workers.

OPTOELECTRONIC AND PHOTONIC SWITCHING: THE TRANSPARENT TELECOMMUNICATION NETWORK

Following the optical fiber transmission systems and light amplifiers that are now part of the national and international telecommunication networks (Chapter 17), several telecommunication laboratories, including British Telecom, are now exploring the possibilities of optical switching at exchanges, with a view to providing faster, more flexible and effective, and ultimately more economical telecommunication services. This field is now wide open to innovation that could drastically change the structure and modes of operation of the network in the longer term.

One approach is to employ optical switches using photons to replace the semiconductor devices using electrons in conventional exchange switch configurations such as crossbar, but with essentially electronic control of the switching function—hence "optoelectronic." Another approach uses guided optical waves in a lithium niobate crystal which can be deviated from one guide to another by applied electrical control signals. This principle can be extended to light beams in space which can be deviated orthogonally to impinge on light-sensitive receptors on a two-dimensional matrix.

Multistage switching networks embodying these principles lend themselves to self-routing capabilities in which packet-type data with appropriate header addresses can find their own way through a complex network, setting up the switches on their own route as they proceed (Chapter 19).

A radically advanced approach to optical switching leads to the concept of light-wavelength multiplexing based on highly coherent laser light sources which can be marginally deviated in frequency to select one of a matrix of frequency-selective light-sensitive receptors. While this technique can be envisaged as applying initially to switching, it leads to a more long-range concept of a completely "transparent" network.

The Transparent Network: The All-Pervasive Optical "Ether"

Proponents of this concept claim that there is sufficient bandwidth available in optical fiber networks, and the frequency stability in coherent light laser sources is potentially sufficient, for each customer in the national network to be allocated a unique frequency. Optical amplification would be used to overcome fiber and distribution losses. In such a system exchange switching would disappear and there would be complete freedom, within defined limits, to use the communication capability of each connection thus established.

For further discussion of the fascinating possibilities of photonics the reader is referred to Ref. 6, which also contains a detailed bibliography of this subject.

INFORMATION SUPERHIGHWAYS AND INTERNET

The volume of information stored in data banks throughout the world is so vast, and is so widely dispersed, that there is a major problem in providing access to the specific items of information that each user may require. And moreover the types of information are themselves diverse—alphanumeric, voice, still picture/graphic, facsimile, or video.

This problem was addressed initially by American academic and university researchers and has resulted in the creation by world corporate telecommunication and computer organizations of the "Internet."

Internet is a worldwide public service communication facility that enables anyone with a computer and the appropriate equipment to connect it via a modem into the world telecommunication network and communicate with any other similarly equipped user. In particular,

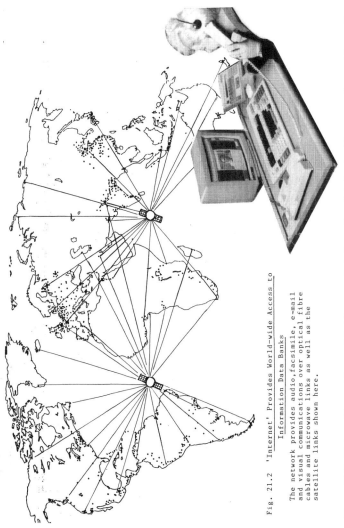

Fig. 21.2 'Internet' Provides World-wide Access to
Information Data Banks

The network provides audio,facsimile, e-mail
and visual communications over optical fibre
cables and microwave links as well as the
satellite links shown here.

Figure 21.2. "Internet" provides worldwide access to information data banks. The network provides audio, facsimile, e-mail, and visual communications over optical fiber cables and microwave links as well as the satellite links shown here.

Internet can be accessed by electronic mail over ordinary telephone circuits as described in Chapter 19. Other multimedia modes that can be transmitted in a telephone circuit bandwidth include person-to-person conferencing, facsimile, and still picture/graphics.

As the volume of worldwide information communication grows, and as wider-bandwidth multimedia services such as high-definition video, group-to-group video conferencing, and fast data-file transfer expand, so the information highways linking cities within a country and across national boundaries will themselves need to grow in capacity. Fortunately the means to provide for this expansion are already available in the ISDN with its internationally agreed technical standards, larger-capacity communication satellites, and optical fiber cables. It is worthy of note that less than 1% of the potential communication capacity of an optical fiber cable has as yet been exploited. Current BT research on a "switched multi-megabit data service" (SMDS) with "asynchronous transfer mode" is recognized as preparing the way for future broadband, high-bit-rate services.[3]

THE IMPACT OF TELECOMMUNICATIONS AND INFORMATION TECHNOLOGY ON THE FUTURE OF CIVILIZATION

But perhaps telecommunications is now moving into an era where technological innovation is relatively less important than innovation in the use of information technology for environmental improvement. It is a simple truism that we no longer need to travel to communicate—there are now available the means to communicate at a distance, person to person, group to group, and with or between computers and data banks, in a variety of modes including aural, visual, and permanent record (fax) forms (Refs. 1, 2, and Chapter 19).

A great deal of the work that now occurs in offices in large cities is concerned with the processing and exchange of information; to perform the work, large numbers of commuters travel by train and car, day after day, from and to their homes in the suburbs or other locations. Such travel wastes time, involves the consumption of vast amounts of irreplaceable fossil fuels, and creates heavy peak demands on road and rail services leading to congestion and delays. At present, information technology is being used on a relatively limited scale to replace travel—mainly by firms with existing offices in cities and factories in other locations.

Long-distance intercontinental travel by jet aircraft for business purposes is wasteful of resources, time-consuming, and destructive of the environment. Teleconferencing offers an economical and speedy alternative that further innovation will improve in communication effectiveness. Furthermore it offers a unique facility for conferring simultaneously between several users in distant and different locations.

What is now needed is a cooperative and constructive approach by government, industry, and environmental groups to determine the benefits to the economy and the environment that could be achieved by the wider use of information technology to minimize travel by enabling offices, factories, and other centers of employment to be moved away from large cities and relocated where they would be environmentally more acceptable. If the benefits are both substantial and achievable—as they may well be—means for encouraging relocation and supporting a location planning policy for new offices and factories would be needed.

The innovative use of information technology could also have a beneficial effect on the rural economy, at present suffering from depopulation and the loss of schools, shops, and other facilities, by stimulating the development of small office and factory units in villages, linked to parent bodies elsewhere. There are possibilities too for "working from home" or in a suitably equipped room in a village hall, i.e., a "telecottage."

A further benefit of the wider application of information technology to the geographical dispersal of office activity could well be to reduce the targets for terrorist bombs such as has occurred in the City of London with its concentrated and massive office blocks in the city center.

But, as noted earlier in this chapter, there is the overriding question as to the impact the world interactive network—the Internet—may have in the longer term on modes of thought, political processes, and even the development of civilization itself.

It is clear that innovators and innovation will have a continuing and important role to play in the future—perhaps with changing emphasis relative to the past—but nevertheless a role that could be of vital importance in determining where and how people live and work, and the quality of life they enjoy, in the years to come.

REFERENCES

1. Murray Laver, *Computers, Communications and Society* (Oxford University Press, London, 1975).
2. Murray Laver, *Information Technology: Agent of Change* (Cambridge University Press, London, 1989).
3. Innovation '94 Symposium at British Telecom Laboratories, Martlesham, Suffolk, 1994; "Innovation '94," *BT Magazine Competitive Edge,* Issue 21, May 1994.
4. John May, "The Shape of Things to Come," *Sunday Telegraph Magazine,* September 1994.
5. R. Smith, "The End of Scientific Journals," Letter, *Br. Med. J., 304,* March 1992.
6. J. E. Midwinter, "Photonics in Switching: The Next Twenty Years of Optical Communications?" Chairman's Address, Electronics Division, Institution of Electrical Engineers, London, October 1990.

22

And Part of Which I Was

Recollections of a Research Engineer

The above is the title of a book by George Brown who played an important role in the evolution of the NTSC color television standards in the United States (Chapter 9)—and it seems a not inappropriate heading under which the present author can look back over a lifetime in telecommunications research and development, which saw the beginnings of most of the major developments described in earlier chapters and in some of which he had a personal involvement. Since it is necessarily an account of events seen through one pair of eyes, perhaps he may be forgiven for using the personal pronoun in this chapter! This account is largely based on my book *Memoirs of a Telecommunications Engineer* (Institution of BT Electrical Engineers, British Telecom HQ, London).

I will begin with schooldays, which in retrospect can now clearly be seen to have set the stage for the career that followed.

A SCHOOLBOY IN THE WORLD OF RADIO

Sound radio broadcasting was just beginning in the United Kingdom in the 1920s when I was a schoolboy at the Portsmouth, Hants, Southern Secondary School. My schoolboy friends and I found ourselves in an exciting new hobby world in which our limited pocket money could buy copper wire and other items enabling us to make crystal detector radio sets that brought, as if by magic and to the amazement of our parents, voices and music from far distant places. It is difficult to realize in this affluent age what a boon to the not-so-well-off it was then to have this ready access, at little cost, to the best of music and the spoken words of the gifted and famous.

And then came further magic in the shape of one's first thermionic valve—a marvelous Aladdin's lamp that enabled the electrical output of a crystal radio set to be amplified to drive a homemade loudspeaker built from a headphone earpiece to which was attached a pleated-paper diaphragm. From one valve there followed more ambitious multivalve sets, with longer range but equally homemade from the instructions in the "wireless" magazines of the time.

The schoolboy's one valve was also pressed into service as an oscillator to make a shortwave transmitter—it was a real thrill to see for the first time enough radio-frequency energy being fed into a half-wave Hertzian dipole to light a small flash-lamp bulb. Equally thrilling, but highly illegal because I lacked a Post Office radio transmitting license, was the ability to communicate by radio over a few miles with other schoolboy friends.

About this time I discovered by chance in a public lending library a remarkable little book called *The Boy Electrician* by Morgan, remarkable because it taught youngsters not only how to construct small electric motors and the like but also to make from scrap materials accessible to a schoolboy electrical measuring instruments such as voltmeters and ammeters, and how to calibrate them from homemade standard cells, Eureka wire resistors, and a Wheatstone bridge. I was fascinated to find, much later in life, that two presidents of the Institution of Electrical Engineers and a distinguished professor of electrical engineering at University College, London, had also "cut their teeth" on *The Boy Electrician*.

THE SCHOOLBOY BECOMES A DOCKYARD
ELECTRICAL ENGINEERING APPRENTICE

These schoolboy explorations in the realm of electronics pointed my footsteps toward an electrical engineering apprenticeship in Portsmouth Naval Dockyard from 1927 to 1933, where the practical training in the Yard workshops and on ships was supported by theoretical studies in the Dockyard School. Life in the Dockyard School was highly competitive; only a fraction of the apprentices in each year survived the end-of-term examinations to continue in the "Upper School" which aimed at creating the professional mechanical and electrical engineers and ship constructors needed for the Admiralty service. The remaining apprentices were destined for the "Lower School" which trained them as technicians, craftsmen, and foremen. The quality of training in the Dockyard Schools was superb, the fourth-year Upper School apprentices reaching engineering degree standards. The numbers of professional engineers thus trained exceeded Admiralty requirements and many found their way into industry either directly or via scholarships such as the Whitworth or Royal scholarships to university.

Incidentally, it was during my apprenticeship in Portsmouth Dockyard that I designed and made from scrap material a moving-coil loudspeaker; it was a source of great satisfaction to find that the measured gap flux density agreed tolerably well with the calculated value. Even more agreeable was to demonstrate to friends the excellent quality of sound reproduction that could be achieved.

Another artifact made during my apprenticeship was the construction of a Baird-type spinning disk 30-line television receiver that actually received in Portsmouth crude images from the London 2LO medium-wave transmitter more than 100 miles away. This was of course long before "real television" began but it was perhaps an indicator of the shape of things to come.

Needless to say, both of these "artifacts" were not recognized parts of the Admiralty training course—they were more the product of an apprentice's fascination with radio and a very tolerant Yard foreman!

THE APPRENTICE ENTERS UNIVERSITY

I left the Naval Dockyard in 1933 with a Royal Scholarship to join the "Light Current" Electrical Engineering course under Professor C. L. Fortescue at City and Guilds Engineering College (part of Imperial College), South Kensington, London. This proved a stimulating introduction to the world of telecommunication technology as it then existed and laid a sound theoretical basis for the future. As Secretary of the College Radio Society it was my good fortune to have enlisted as a speaker one Robert Watson-Watt (later Sir Robert) of the U.K. Radio Research Station, Slough, who talked about the location of thunderstorms by radio—an interesting precursor to his pioneering work on radar that had a dominating influence on the outcome of the Battle of Britain in World War II.

HOW "SIDEBANDS" PLAYED A PART IN DECIDING A CAREER

In 1935, equipped with B.Sc. and M.Sc.(Eng.) degrees from Guilds, I entered the Open Competition for a post as Assistant Engineer in the Post Office Engineering Department. This included an interview with the Chief Engineer, Sir Archibald Gill—an interview during which he asked: "Do you believe in the objective existence of sidebands (i.e., of a modulated carrier wave)?" It may seem remarkable that such a question should even have been posed, but at the time there was still a lingering controversy, mainly in academic circles, as to whether sidebands were just a convenient mathematical fiction or whether they were "real." Luckily I had witnessed at Guilds a convincing demonstration involving a frequency-swept tuned circuit and a sine-wave modulated carrier, the response being displayed on a primitive electromechanical Dudell oscilloscope and revealed as a triple-peaked curve. So my response to the question was a triumphant "Yes, I have seen them!"

This must have "done the trick" for an offer of a job as an Assistant Engineer in the PO Engineering Department soon came—coincidentally with an offer from the Admiralty of an Electrical Engineer Cadetship at the Royal Naval College, Greenwich. But the possibility of working on radio with the PO won the day and I happily accepted the former offer.

Thus began a career, initially with the PO Radio Experimental Branch at Dollis Hill, London, which spanned more than 40 years and saw at firsthand the remarkable technological evolution that has created today's worldwide multifaceted telecommunication services. To complement the formal account of this evolution described in earlier chapters, perhaps some personal reminiscences may be of interest.

LONG-WAVE RADIO: A DAUNTING TASK FOR A YOUNG ENGINEER

The Post Office Engineering Department in which I found myself in 1933 had designed and built one of the most powerful radiotelegraph transmitting stations in the world—GBR at Rugby, England, which provided worldwide telegraphic news and time signal service. GBR operated on the incredibly low radio frequency of 16 kHz, corresponding to a wavelength of 18,750 meters. The sixteen 820-ft-high masts towered over the countryside for more than a mile, and a bank of large water-cooled valves fed a current of some 1000 amperes into the aerial.

Imagine the feelings of a very young engineer when required to measure the impedance of this vast aerial system with a "tuned circuit substitution" measuring set which looked minute compared with the 12-ft-diameter aerial upload, which required a gang to move it to the measuring set. But the mechanical problems of scale faded into insignificance when I realized that the GBR aerial was a highly efficient collector of energy from thunderstorms far out into the Atlantic or over Siberia—and thousands of volts appeared at times between the aerial and earth. That I survived this hazard can be attributed to a farsighted colleague who designed a relay automatically to earth the aerial when the voltage became excessive—otherwise this book might never have been written!

SHORTWAVE RADIO: THOSE SIDEBANDS AGAIN

From 1933 to the outbreak of the Second World War in 1939 my activities at the PO Radio Experimental Branch at Dollis Hill were centered mainly on the design of receivers and transmitter drives to convert the PO overseas shortwave radiotelephone services from double-sideband to single-sideband operation, which substantially improved quality and made possible four speech channels in each frequency allocation. Since the early work on single-sideband operation was carried out in cooperation with Bell Laboratories and AT&T engineers over the transatlantic radiotelephone service, this brought several interesting American contacts, and the names on technical papers soon became real people and in some cases personal friends.

All this seemed a very logical extension of the PO Engineer-in-Chief's question as to whether I believed in the objective existence of sidebands. But my dominant feeling at the time was, "How marvelous—I am actually being paid to do what I had hitherto enjoyed only as a hobby!"

THE MOST COMPLEX SHORTWAVE
RECEIVING SYSTEM EVER DEVISED

My involvement with Bell Laboratories continued with the design and building of the Multiple-Unit Steerable Antenna (MUSA; Chapter 7). With its 16 in-line rhombic aerials on the flat Cooling Marshes of the Thames Estuary, and an elaborate phasing system to combine the signals from the aerials, the MUSA was the most complex shortwave receiving system ever devised. It represented a last bid to improve the shortwave transatlantic radio telephone service, before the advent of submarine cables and satellites; it did valuable service in the war years, for example enabling Churchill and Roosevelt to talk directly to one another.

For my colleagues and I there was an amusing spin-off—the local inhabitants of the Cooling Marshes were convinced that we were building a "death ray" that would bring down German bombers as they swept up the Thames Estuary toward London. We did not disillusion them; after all, who would wish to forego the prestige, not to say the free pints of beer, that the locals offered!

A NEARLY LETHAL SHORTWAVE DIRECTION FINDER

My wartime activities with the PO involved, among other matters, setting up and calibrating shortwave radio-direction finding stations for the Admiralty—vital for tracing the movements of the German fleet in and around the North Sea. One such station at the PO Radio Station at Baldock, Hertfordshire, was unique in that the receiving equipment was installed in an underground steel tank. After a long calibrating session I had occasion to sleep in the tank on a cold winter night—totally forgetting that CO_2 gas is not only heavier than air but is potentially lethal. Fortunately the outcome the morning after was no worse than the "mother of all" headaches, otherwise this book, once more, might never have been written!

THE MOON AS A COMMUNICATION SATELLITE; OR WAS IT "MOONSHINE"?

In 1940, when the Battle of Britain was raging and a handful of RAF fighter pilots were deciding the fate of the Western world, I began to speculate on the possibility of using the moon as a passive communication satellite. Radio waves scattered from its surface might be used to provide communication between any two points on the Earth's surface when the moon was simultaneously visible to both.

In order to concentrate as much radio wave energy as possible on the moon, which subtends an angle of about 0.5° at the Earth, microwaves, i.e., radio waves of some 10-cm wavelength, and a 10-meter-diameter dish aerial would be suitable. With a microwave power of about 1 kW, calculation showed that a weak but detectable signal should be received. Furthermore, consideration of the nature of the scattering from the rough surface of the moon indicated a time-spread of the received signal of about 1 millisecond; this would not be a barrier to satisfactory voice communication and would not adversely affect telegraphy. However, the distance of the moon from the Earth, a quarter of a million miles, meant that there would be a delay of at least 2.5 seconds between question and answer.

Visualizing that a moon reflection link might have military, if not civilian, applications I prepared a memorandum on the possibilities for the higher levels of the PO Engineering Department; however, it got rather a "dusty" answer: did I not realize that "there was a war on" and I should not be "wasting my time on moonshine"!

Sir Edward Appleton in his classic paper on "The Scientific Principles of Radio-Location," presented at the Institution of Electrical Engineers, London, in 1946, revealed that the U.S. Army Signal Corps had in 1945 received signals scattered from the moon. And in 1959 scientists at Bell Laboratories bounced microwave signals from the moon as a precursor to tests with their Earth-orbiting balloon satellite "ECHO."

TELEVISION: BEATING A SWORD INTO A PLOUGHSHARE

The British Broadcasting Corporation's television service at Alexandra Palace, London—which opened in 1936 as the first public high-definition television service in the world—was closed down at the outbreak of war in 1939 to avoid giving guidance to German bombers. The closure was appropriately in the middle of a Mickey Mouse film with the words "I t'ink I go home!" When the service was resumed after the war, it was started at the same point in the film as it had closed.

I had resolved to see this beginning and, since commercial television receivers were few and expensive, decided to make my own. As a member of a Ministry of Defence "Panel for Allocating Captured German Electronic Equipment," I had allocated myself a German radar set. This, with a British radar IF strip, fortunately tuned to 45 MHz (the frequency of the BBC Alexandra Palace vision transmitter), enabled me to construct on the kitchen table without benefit of test equipment other than an "Avometer," a workable television receiver.

True the picture was small and green in color, and the picture linearity could have been better, but it was live television and friends came from miles around to see this marvel. And, in a sense, there was satisfaction in having "beaten a German sword into a British ploughshare!"

TELEVISION PIRATES: THE POST-MASTER GENERAL'S DILEMMA

Television broadcasting in the United Kingdom developed on the basis of a license paid by the owners of television receivers and collected by the Post Office, from which the British Broadcasting Corporation was funded. As the numbers of television receivers grew into millions in the 1950s, a serious problem arose because many viewers avoided paying for a license—in fact they became "television pirates." In an attempt to scare the pirates into paying up, the then Post-Master General, Earl De La Warr, had advised Parliament that users of unlicensed television sets could be located and possibly prosecuted. Unfortunately at the time no such technical means existed—creating a situation of some embarrassment to the PMG and the Post Office.

The PMG paid a visit to Dollis Hill to discuss the problem and I was brought in to give such technical advice as could be offered. It occurred to me that the line-scanning coils of a working television receiver produced a magnetic field at the line-scanning frequency of 10 kHz that could extend some distance from the receiver and pass through brick walls. Furthermore the sawtooth waveform was rich in harmonics, with a strong second harmonic at 20 kHz. Thus, we were able to demonstrate to the PMG that a radio receiver tuned to 20 kHz and equipped with a loop aerial could in fact pick up a signal—heard as a "whistle"—up to several yards from a working television receiver.

From this crude demonstration at Dollis Hill a Post Office "Television Detection Van" was quickly developed; with three very visible loop aerials on its roof it was possible by comparing the signal strength from each of the aerials to determine whether a television receiver was to the right or left of, in front of or behind, the van. That it worked was publicized by the BBC and eventually the mere presence of the TV Detection Van in a district was enough to stimulate a rush of new license holders. During the years the vans were in operation, they must have stimulated many millions of pounds of additional license revenue—I only wish I had had the foresight to claim a mere 5% royalty on this revenue!

ONWARDS IN THE RADIO SPECTRUM FROM SHORTWAVES TO MICROWAVES

It seemed a logical step to progress from an involvement in the development and use of shortwaves to similar activities in the microwave field, the history of which has been outlined in Chapter 11. The fascination of microwaves lay in the fact that they could be accurately beamed by dish aerials of convenient size and accommodate thousands of telephone circuits or a high-quality television channel on each radio carrier. Furthermore, since microwaves could "hop" line-of-sight from hilltop to hilltop, links could be established relatively rapidly compared with coaxial cables laid in the ground and requiring a multiplicity of "wayleaves."

These considerations led to an ongoing battle between the "Lines Branch" of the PO Engineering Department, which had a strong bias toward coaxial cables, and the "Inland Radio Branch" of which I was then Head. Lines Branch of course made much of the susceptibility of microwaves to "fade" under adverse propagation conditions. Eventually it was agreed that wholly acceptable standards of transmission could be established by appropriate design using microwaves.

I had much admired the pioneering work on microwaves carried out by Bell Laboratories in the 1950s, notably at the Holmdel Laboratory under H. T. Friis. It will be recalled (from Chapter 11) that when I asked him how big he thought an ideal research laboratory should be, he said, "Perhaps about 100 people, I could then handpick them for the jobs to be done"—a somewhat surprising comment when research staff are now often measured in thousands.

From the work on microwaves carried out by the PO Radio Experimental Branch at Dollis Hill, stimulated by my contacts with the Bell Laboratories at Holmdel, and in cooperation with U.K. industry, notably Standard Telephones and Cables Ltd. and the General Electric Company, there developed a U.K. national microwave network with its focus in the PO, now BT, Tower in London. This network carried most of the intercity television links in the United Kingdom and large numbers of telephone circuits.

If I can claim a personal contribution it was an early recognition of the important role the traveling-wave tube microwave amplifier would play in this development, and later in satellite communications. It came about from a friendship with Rudolf Kompfner, the Austrian architect who invented the traveling-wave tube at the Clarendon Laboratory, Oxford University, during the war years (Chapter 11).

MICROWAVES BECOME "RESPECTABLE" AND ARE STANDARDIZED INTERNATIONALLY

Working in the microwave field led to my participation during the 1950s in the work of the International Radio Consultative Committee (CCIR) of the International Telecommunication Union, Geneva, as one of a U.K. delegation led by an Assistant Engineer-in-Chief of the Post Office Engineering Department, Captain C. F. Booth. My particular responsibility was in the standardization of the technical characteristics and transmission standards of microwave links to facilitate country-to-country connections.

This proved a most rewarding and educational experience—initially in cooperating with U.K. telecommunications industry representatives in the preparation of agreed U.K. proposals. These had then to be discussed with other countries at meetings of the CCIR Study Group (IX) on Microwave Radio-Relay Systems, and put into a form suitable for presentation at plenary meetings of the CCIR. I was, for a time, International Chairman of Study Group IX, an experience that taught me a lot about the gentle art of compromise without giving too much away on desirable U.K. objectives!

Contact with members of the CCIR Study Group from other countries and working toward common objectives brought many friendships that continued in later years, and helped also in new fields such as the early development of satellite communications. But it was not all hard work in the study groups and plenaries; meetings were held at times in different countries and the host country traditionally offered hospitality in dinners and social visits of a moderately generous character. I recall in particular a plenary meeting in Los Angeles where AT&T presented all delegates with the freedom of Catalina Island, which included free meals and drinks in any bar!

The value of international cooperation through the ITU and its Consultative Committees (CCIR and CCITT) lay not only in the technical and operating standards these organizations defined, it was the demonstration that even in a world divided at the time by the Iron Curtain it was possible to achieve technical agreements between engineers that transcended national boundaries and ideologies.

A SABBATICAL YEAR IN THE UNITED STATES—AND ANGLO-AMERICAN FRIENDSHIPS ARE FORMED

It was my good fortune in 1955 to be awarded a Commonwealth Fund Fellowship for a year's "advanced study and travel" in the United States. This proved to be another rewarding experience, not only for the knowledge gained but especially for the friendships and contacts it established with people engaged in similar professional activities, contacts that later had an important bearing on joint transatlantic projects such as the AT&T/Bell and British Post Office involvement in Project TELSTAR.

The CFF was established in 1925 by the wealthy Harkness family with the aims of supporting medical research and educational projects and promoting better understanding between the United States and British Commonwealth and European countries. To the latter end CFF endowed some 20 traveling fellowships in a wide range of categories that included science, engineering, medicine, business studies, law, literature, and journalism. In the field of journalism a Commonwealth Fund Fellow who made very effective use of his studies was Alistair Cooke, who became a remarkably successful unofficial ambassador for the United States via his broadcast talks "Letter From America."

A very tolerant Director of the Commonwealth Fund allowed me to make my own program which included visits to, and discussions with, people at the following locations:

American Telephone and Telegraph Co., New York
Bell Telephone Laboratories, Murray Hill and Holmdel, New Jersey
National Bureau of Standards, Colorado
Federal Communications Commission, Washington, D.C.
Federal Telecommunication Laboratories, Nutley, New Jersey
Radio Corporation of America, Princeton, New Jersey
Massachusetts Institute of Technology, Cambridge

While most of these visits were concerned with studying advanced telecommunication technology with research scientists and engineers, others such as the visit to the Federal Communications Commission were concerned with the management and control of the expanding use of the radio spectrum by sound and television broadcasting, point-to-point microwave radio-relay systems, and mobile radio communication. A valuable feature of such discussions was that they often provided a foreknowledge of similar problems that were to become significant a few years later on the other side of the Atlantic.

But it was not all study and hard work. A generous CFF encouraged its Fellows, for part of their time, to travel widely in the United States, not only visiting universities but also seeing something of the scenic grandeur of the country. And to this end they provided not only enough cash to acquire a secondhand car, but encouraged the wives of Fellows to join them on their travels.

And so it was that my wife and I traveled some 12,000 miles around the United States in 3 months, starting in a New Jersey snowstorm in late April, going south to the spring in Florida, along the Gulf Coast and through Texas and Arizona, over the Coast Range of mountains to San Diego, and up the Pacific Coast to San Francisco. Then eastward over the Rocky Mountains to Yellowstone Park—where we were back in the snows of winter!—through the Midwest and on to New York and, eventually, New Jersey.

It was an illuminating experience, finding a warm welcome from Bell System managers and professorial staff in universities wherever we went. But perhaps the most valuable aspect of the fellowship was the insight it gave me into the organization and motivation of the Bell System and its Laboratories, and its role in creating what was universally acknowledged as the most effective telecommunication system then in existence.

SATELLITES—A DREAM COME TRUE, AND SOME ODD CHANCES ALONG THE WAY

The early history of satellite communication has been told in Chapter 15, and my speculation in 1940 on the possibility of using the moon as a passive communication satellite has been mentioned earlier in this chapter.

The Russian Sputnik of 1957 and the launch of the AT&T/Bell

Project TELSTAR in 1962 were undoubtedly the keys to A. C. Clarke's dream of 1945 coming true.

My involvement began as Head of the newly formed "Space Communication Systems" Branch of the Post Office Engineering Department Headquarters in London, created by Engineer-in-Chief Sir Albert Mumford at a time when I am quite sure the highly conservative PO Administration felt that what "those engineers" were committing them to was pure science fiction!

My good friend and colleague John Taylor was then Head of the Radio Experimental Branch at Dollis Hill, which designed and built much of the key equipment for the first satellite earth station at Goonhilly, Cornwall. By great good fortune we happened to be next-door neighbors, living at Wembley Park, NW London. Never was there closer collaboration between a Headquarters and a Dollis Hill Branch, often with decisions arrived at and sealed with a glass of Scotch whisky on a Sunday morning!

Another fortuitous accident put us on track for an aerial designer with the unique skills needed to build a large satellite earth station aerial in less than one year. My daughter, then at the North London Collegiate School, mentioned to a school friend that her father wanted to build a large aerial. The friend mentioned this to her father in the then Ministry of Supply who telephoned me to ask, "Do you know a Tom Husband who designed the 240-ft-diameter aerial at the Jodrell Bank Radio Astronomy Centre?" Thus began a collaboration that produced within a year an 85-ft-diameter, massive 870-ton, but precisely steerable, aerial for the satellite earth station at Goonhilly.

There were other happy coincidences: the discovery that the Mullard Laboratories had developed a low-noise maser receiving amplifier, and the Services Electronic Research Laboratory could offer a high-power 1-kW traveling-wave tube transmitting amplifier (my friend the traveling-wave tube inventor Rudolf Kompfner must have enjoyed this!). But above all there was the feeling among all concerned that they were participating in a great step forward in world telecommunications—and it was this that enabled Goonhilly to be built from a virgin site on the Lizard Peninsula in Cornwall in one year from July 1, 1961, to July 1, 1962.

The atmosphere at Goonhilly on July 10, 1962, was electric. Would the aerial with its 0.1° beamwidth track accurately a rapidly moving satellite (this was before the days of "stationary" satellites), would all of the complex and virtually untried electronic equipment work satisfactorily?

And all of this had to be done with the eyes of the world, as seen by Raymond Baxter for the BBC and Ian Trethowan for the Independent Broadcasters, watching and reporting our endeavors at Goonhilly— truly a "tightrope walk" situation.

In the event, and after a minor hiatus concerning an Anglo-American misunderstanding about the right- or left-handed polarization of the circularly polarized waves from the satellite (see Chapter 10), all went well and the United Kingdom transmitted the first-ever live television pictures to a viewing audience of millions in the United States. From this we went on, with the help of the BBC Research Department, to make the first-ever color television transmissions to the United States using American NTSC standards. And all of this was achieved at a fraction of the cost of the French and American earth stations, and which, because of the absence of a radome in the PO design, outperformed them in rain.

The Independent Broadcasters had provided a crate of the best French champagne, and on the night of July 11, 1962, after the success of the first-ever both-way television transmission across the Atlantic, the only fair comment is that Goonhilly was airborne!

FROM MICROWAVES AND SATELLITES TO DIRECTOR OF RESEARCH

With my appointment as Director of Research of the Post Office in 1966 and until my retirement in 1975, the joys of hands-on participation in the development of new technology largely disappeared and other, more challenging, responsibilities took over. It was a time when transistor and microchip technology was advancing rapidly, optical fibers were looking more promising, exchange switching was evolving into a computer-controlled electronic form, and, above all, the digital revolution was beginning.

Fortunately I was supported by three Deputy Directors, each an expert in his field, who patiently kept their Director "on the rails"! I have to thank, in particular, Dr. J. R. Tillman for enlightening me on the physics of the transistor and optical fibers, H. B. Law for unraveling the mysteries of digital switching, and E. W. Ayers for guidance in visual systems such as Viewdata/Prestel.

I discovered on a visit to Bell Laboratories that Julius Molnar, Deputy President of Bell, had, like me, been primarily a transmission engineer and had to "learn switching" late in his career, an operation that he described to me as akin to "papering over the cracks"!

As Director I had also to oversee the planning and building of new laboratories at Martlesham, Suffolk, and the movement of some 1500 staff from Dollis Hill, NW London, to the new location. The new laboratories were formally opened by Her Majesty the Queen in 1975; one of her first questions was, "And how did the wives of the staff react to the change from life in London to life in Suffolk?" It was a pleasure to respond that it meant a better quality of life, the opportunity to enjoy pleasant rural surroundings with good local schools for the children, and far less time spent in traveling to and from work by husbands.

Retirement brought an opportunity to act as visiting professor at my old university, Imperial College, and at University College, London. I was also invited to give the Bernard Price Memorial Lecture in South Africa at the universities of Witwatersrand, Cape Town, and Durban—my subject being "Evolutionary Telecommunications and Ecological Man," a preview of the impact of information technology and enhanced telecommunications on the way people may live and work in the future (discussed in Chapter 21).

Mine has been an immensely rewarding professional career, one in which I feel privileged to play a small part in building today's world telecommunication systems and the information technology that may yet help to create a better-ordered, more prosperous and peaceful world. From a schoolboy building a crystal radio set to an engineer helping to span the Atlantic with live television for the first time ever—what a marvelous journey it has been!

Name Index

Subject Index